Constants of Proportionality

" The Heart of The Equation "

Edited by

Paul F. Kisak

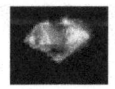

Virginia, USA

Printed in The United States of America

First Trade Edition: 2015
10 9 8 7 6 5 4 3 2 1

Black & White on White paper
207 pages

ISBN-13: 978-1522754428
ISBN-10: 1522754423

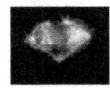

Virginia, USA

Contents

1 Proportionality (mathematics)   **1**

  1.1   Symbols 2

  1.2   Direct proportionality 2

      1.2.1   Examples 2

      1.2.2   Properties 3

  1.3   Inverse proportionality 3

  1.4   Hyperbolic coordinates 3

  1.5   Exponential and logarithmic proportionality 4

  1.6   See also 5

      1.6.1   Growth 5

2 Function (mathematics)   **6**

  2.1   Introduction and examples 6

  2.2   Definition 10

  2.3   Notation 11

  2.4   Specifying a function 12

      2.4.1   Graph 12

      2.4.2   Formulas and algorithms 12

      2.4.3   Computability 12

  2.5   Basic properties 13

      2.5.1   Image and preimage 13

      2.5.2   Injective and surjective functions 14

      2.5.3   Function composition 14

      2.5.4   Identity function 15

      2.5.5   Restrictions and extensions 15

      2.5.6   Inverse function 15

  2.6   Types of functions 16

      2.6.1   Real-valued functions 16

      2.6.2   Further types of functions 16

  2.7   Function spaces 17

	2.7.1	Currying	17
2.8		Variants and generalizations	17
	2.8.1	Alternative definition of a function	17
	2.8.2	Partial and multi-valued functions	18
	2.8.3	Functions with multiple inputs and outputs	18
	2.8.4	Functors	18
2.9		History	19
2.10		See also	19
2.11		Notes	19
2.12		References	20
2.13		Further reading	20
2.14		External links	21

3 Product (mathematics)　　**24**

3.1		Product of two numbers	24
	3.1.1	Product of two natural numbers	24
	3.1.2	Product of two integers	24
	3.1.3	Product of two fractions	25
	3.1.4	Product of two real numbers	25
	3.1.5	Product of two complex numbers	26
	3.1.6	Product of two quaternions	27
3.2		Product of sequences	27
3.3		Commutative rings	27
	3.3.1	Residue classes of integers	27
	3.3.2	Rings of functions	27
	3.3.3	Convolution	28
	3.3.4	Polynomial rings	28
3.4		Products in linear algebra	29
	3.4.1	Scalar multiplication	29
	3.4.2	Scalar product	29
	3.4.3	Cross product in 3-dimensional space	29
	3.4.4	Composition of linear mappings	30
	3.4.5	Product of two matrices	30
	3.4.6	Composition of linear functions as matrix product	31
	3.4.7	Tensor product of vector spaces	31
	3.4.8	The class of all objects with a tensor product	31
	3.4.9	Other products in linear algebra	31
3.5		Cartesian product	32
3.6		Empty product	32

3.7   Products over other algebraic structures 32

3.8   Products in category theory 33

3.9   Other products 33

3.10 See also 33

3.11 Notes 33

3.12 References 33

3.13 External links 33

4   Variable (mathematics)   **34**

4.1   Etymology 34

4.2   Genesis and evolution of the concept 34

4.3   Specific kinds of variables 35

      4.3.1   Dependent and independent variables 36

      4.3.2   Examples 36

4.4   Notation 36

4.5   See also 38

4.6   Bibliography 38

4.7   References 38

5   Coefficient   **40**

5.1   Linear algebra 40

5.2   Examples of physical coefficients 41

5.3   See also 41

5.4   References 42

6   Ratio   **43**

6.1   Notation and terminology 44

6.2   History and etymology 44

      6.2.1   Euclid's definitions 45

6.3   Number of terms and use of fractions 46

6.4   Proportions and percentage ratios 46

6.5   Reduction 46

6.6   Irrational ratios 47

6.7   Odds 47

6.8   Units 47

6.9   Triangular coordinates 47

6.10 See also 48

6.11 References 48

6.12 Further reading 49

6.13 External links 49

7 Correlation and dependence **50**

7.1 Pearson's product-moment coefficient 50

7.2 Rank correlation coefficients 52

7.3 Other measures of dependence among random variables 52

7.4 Sensitivity to the data distribution 53

7.5 Correlation matrices 54

7.6 Common misconceptions 54

    7.6.1 Correlation and causality 54

    7.6.2 Correlation and linearity 54

7.7 Bivariate normal distribution 55

7.8 Partial correlation 56

7.9 See also 56

7.10 References 57

7.11 Further reading 58

7.12 External links 58

8 Dependent and independent variables **59**

8.1 Use 59

    8.1.1 Mathematics 59

    8.1.2 Statistics 59

    8.1.3 Modelling 60

    8.1.4 Simulation 61

8.2 Statistics synonyms 61

8.3 Other variables 61

8.4 Examples 62

8.5 See also 62

8.6 References 63

9 Cartesian coordinate system **64**

9.1 History 64

9.2 Description 65

    9.2.1 One dimension 65

    9.2.2 Two dimensions 67

    9.2.3 Three dimensions 67

    9.2.4 Higher dimensions 67

    9.2.5 Generalizations 67

9.3 Notations and conventions 68

9.3.1 Quadrants and octants 70

9.4 Cartesian formulae for the plane 71

9.4.1 Distance between two points 71

9.4.2 Euclidean transformations 71

9.5 Orientation and handedness 74

9.5.1 In two dimensions 74

9.5.2 In three dimensions 75

9.6 Representing a vector in the standard basis 76

9.7 Applications 77

9.8 See also 78

9.9 Notes 78

9.10 References 78

9.11 Sources 78

9.12 Further reading 79

9.13 External links 79

10 Hyperbolic coordinates **80**

10.1 Alternative quadrant metric 80

10.2 Applications in physical science 82

10.3 Statistical applications 82

10.4 Economic applications 82

10.5 History 83

10.6 References 83

11 Hyperbola **84**

11.1 History 84

11.2 Nomenclature and features 85

11.3 Mathematical definitions 86

11.3.1 Conic section 86

11.3.2 Difference of distances to foci 86

11.3.3 Directrix and focus 86

11.3.4 Reciprocation of a circle 87

11.3.5 Quadratic equation 87

11.4 True anomaly 90

11.5 Geometrical constructions 90

11.6 Reflections and tangent lines 91

11.7 Hyperbolic functions and equations 91

11.8 Relation to other conic sections 92

11.9 Conic section analysis of the hyperbolic appearance of circles 92

11.10 Derived curves 94

11.11 Coordinate systems 94

    11.11.1 Cartesian coordinates 94

    11.11.2 Polar coordinates 95

    11.11.3 Parametric equations 96

    11.11.4 Elliptic coordinates 96

11.12 Rectangular hyperbola 96

11.13 Other properties of hyperbolas 97

11.14 Applications 97

    11.14.1 Sundials 97

    11.14.2 Multilateration 98

    11.14.3 Path followed by a particle 98

    11.14.4 Korteweg-de Vries equation 98

    11.14.5 Angle trisection 98

    11.14.6 Efficient portfolio frontier 98

11.15 Extensions 99

11.16 See also 99

    11.16.1 Other conic sections 99

    11.16.2 Other related topics 99

11.17 Notes 99

11.18 References 100

11.19 External links 100

12 Hyperbolic growth   **115**

12.1 Description 115

    12.1.1 Comparisons with other growth 115

12.2 Applications 115

    12.2.1 Population 115

    12.2.2 Queuing theory 116

    12.2.3 Enzyme kinetics 117

12.3 Mathematical example 117

12.4 See also 117

    12.4.1 Mathematics 117

    12.4.2 Growth 118

12.5 References 118

12.6 References 118

13 Multiplicative inverse   **119**

13.1 Examples and counterexamples 120

13.2 Complex numbers 120

13.3 Calculus 121

13.4 Algorithms 122

13.5 Reciprocals of irrational numbers 123

13.6 Further remarks 123

13.7 Applications 124

13.8 See also 124

13.9 Notes 124

13.10References 124

14 Linear function   **125**

14.1 As a polynomial function 125

14.2 As a linear map 125

14.3 See also 127

14.4 Notes 127

14.5 References 128

14.6 External links 128

15 Exponential function   **129**

15.1 Formal definition 130

15.2 Overview 130

15.3 Derivatives and differential equations 133

15.4 Continued fractions for e^x 134

15.5 Complex plane 135

    15.5.1 Computation of a^b where both a and b are complex 136

15.6 Matrices and Banach algebras 136

15.7 Lie algebras 136

15.8 Double exponential function 136

15.9 Similar properties of e and the function e^z 137

15.10See also 137

15.11References 137

15.12External links 138

16 Logarithm   **141**

16.1 Motivation and definition 142

    16.1.1 Exponentiation 143

    16.1.2 Definition 143

    16.1.3 Examples 143

16.2 Logarithmic identities 143

16.2.1 Product, quotient, power and root 144

16.2.2 Change of base 144

16.3 Particular bases 144

16.4 History 145

16.4.1 Predecessors 145

16.4.2 From Napier to Euler 145

16.4.3 Logarithm tables, slide rules, and historical applications 148

16.5 Analytic properties 150

16.5.1 Logarithmic function 150

16.5.2 Inverse function 151

16.5.3 Derivative and antiderivative 151

16.5.4 Integral representation of the natural logarithm 153

16.5.5 Transcendence of the logarithm 155

16.6 Calculation 155

16.6.1 Power series 156

16.6.2 Arithmetic–geometric mean approximation 157

16.7 Applications 158

16.7.1 Logarithmic scale 159

16.7.2 Psychology 160

16.7.3 Probability theory and statistics 160

16.7.4 Computational complexity 161

16.7.5 Entropy and chaos 162

16.7.6 Fractals 163

16.7.7 Music 163

16.7.8 Number theory 164

16.8 Generalizations 164

16.8.1 Complex logarithm 164

16.8.2 Inverses of other exponential functions 167

16.8.3 Related concepts 167

16.9 See also 167

16.10 Notes 168

16.11 References 168

16.12 External links 173

17 Ceteris paribus   **174**

17.1 Economics 174

17.1.1 Clause 174

17.1.2 Characterization given by Alfred Marshall 175

17.1.3 Two uses 175

17.2 See also 175

17.3 References 176

17.4 External links 176

18 Scientific modelling **177**

18.1 Overview 178

18.2 Basics of scientific modelling 178

18.2.1 Modelling as a substitute for direct measurement and experimentation 178

18.2.2 Simulation 178

18.2.3 Structure 179

18.2.4 Systems 179

18.2.5 Generating a model 179

18.2.6 Evaluating a model 179

18.2.7 Visualization 180

18.3 Types of scientific modelling 180

18.4 Applications 180

18.4.1 Modelling and Simulation 180

18.4.2 Model Based Learning in Education 180

18.5 See also 182

18.6 References 183

18.7 Further reading 183

18.8 External links 184

18.9 Text and image sources, contributors, and licenses 185

18.9.1 Text 185

18.9.2 Images 191

18.9.3 Content license 195

Chapter 1

Proportionality (mathematics)

For other uses, see Proportionality.

In mathematics, two variables are **proportional** if a change in one is always accompanied by a change in the other, and

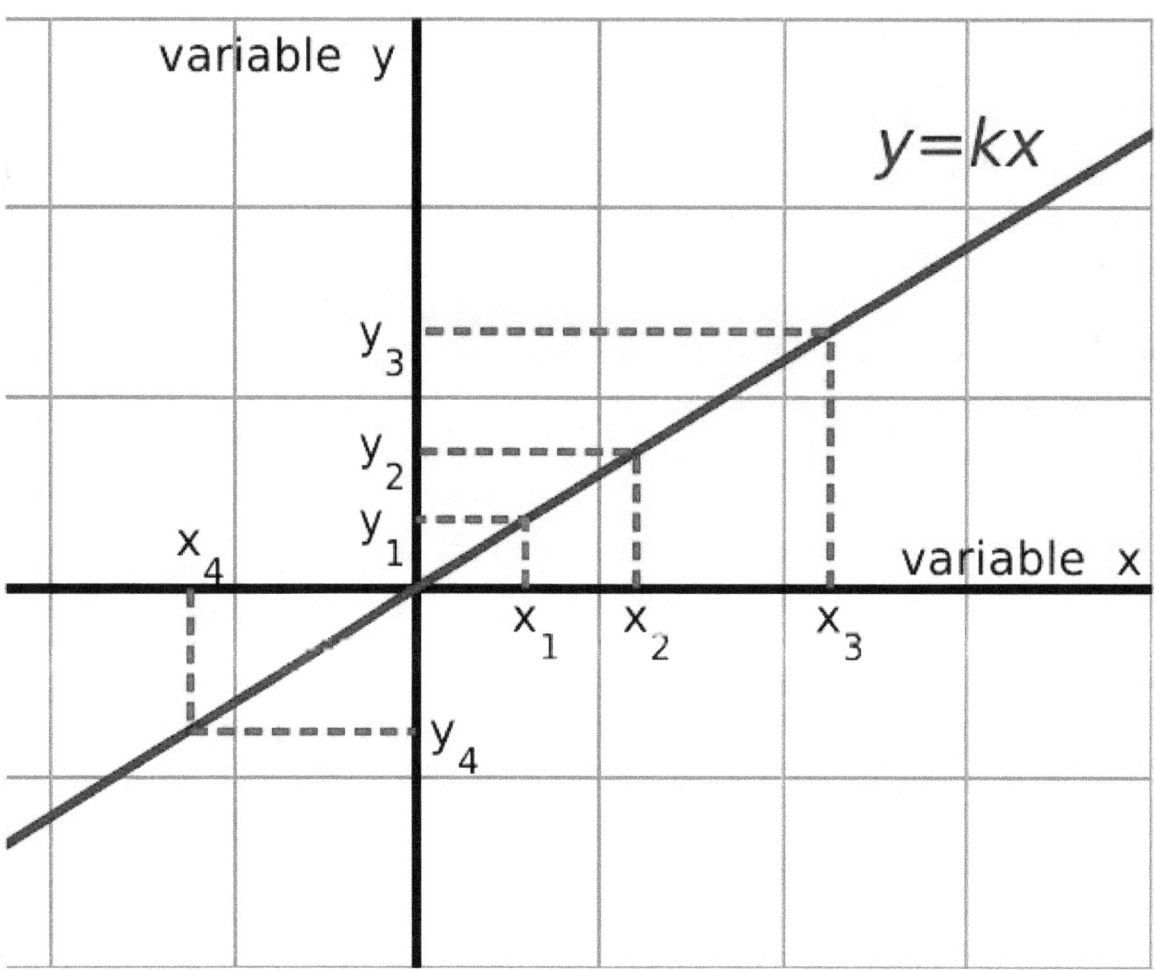

Variable y is directly proportional to the variable x.

if the changes are always related by use of a constant multiplier. The constant is called the coefficient of proportionality or **proportionality constant**.

- If one variable is always the product of the other and a constant, the two are said to be *directly proportional*. x and y are directly proportional if the ratio $\frac{y}{x}$ is constant.

- If the product of the two variables is always equal to a constant, the two are said to be *inversely proportional*. x and y are inversely proportional if the product xy is constant.

In order to express the statement "y is (directly) proportional to x" mathematically, we write an equation $y = cx$, for some real constant, c. Symbolically, this is written $y \propto x$.

In order to express the statement "y is inversely proportional to x" mathematically, we write an equation $y = c/x$. We can equivalently write "y is proportional to $1/x$", which $y = c/x$ would represent.

If a linear function transforms 0, a and b into 0, c and d, and if the product $a\ b\ c\ d$ is not zero, we say a and b are proportional to c and d. An equality of two ratios such as $\frac{a}{c} = \frac{b}{d}$, where no term is zero, is called a proportion.

1.1   Symbols

The mathematical symbol $\propto$ (U+221D in Unicode, \propto in TeX) is used to indicate that two values are proportional. For example, A $\propto$ B means the variable A is directly proportional to the variable B.

Other symbols include:

- ∷ - U+2237 "PROPORTION"

- ∺ - U+223A "GEOMETRIC PROPORTION"

1.2   Direct proportionality

Given two variables x and y, y *is **directly proportional** to* x (*x and y **vary directly**, or x and y are in **direct variation***) if there is a non-zero constant k such that

$$y = kx.$$

The relation is often denoted, using the $\propto$ symbol, as

$$y \propto x$$

and the constant ratio

$$k = \frac{y}{x}$$

is called the **proportionality constant, constant of variation** or **constant of proportionality**.

1.2.1   Examples

- If an object travels at a constant speed, then the distance traveled is directly proportional to the time spent traveling, with the speed being the constant of proportionality.

- The circumference of a circle is directly proportional to its diameter, with the constant of proportionality equal to π.

- On a map drawn to scale, the distance between any two points on the map is directly proportional to the distance between the two locations that the points represent, with the constant of proportionality being the scale of the map.

- The force acting on a certain object due to gravity is directly proportional to the object's mass; the constant of proportionality between the mass and the force is known as gravitational acceleration.

1.2.2 Properties

Since

$$y = kx$$

is equivalent to

$$x = \left(\frac{1}{k}\right) y,$$

it follows that if y is directly proportional to x, with (nonzero) proportionality constant k, then x is also directly proportional to y with proportionality constant $1/k$.

If y is directly proportional to x, then the graph of y as a function of x will be a straight line passing through the origin with the slope of the line equal to the constant of proportionality: it corresponds to linear growth.

1.3 Inverse proportionality

The concept of *inverse proportionality* can be contrasted against *direct proportionality*. Consider two variables said to be "inversely proportional" to each other. If all other variables are held constant, the magnitude or absolute value of one inversely proportional variable will decrease if the other variable increases, while their product (the constant of proportionality k) is always the same.

Formally, two variables are **inversely proportional** (also called **varying inversely**, in **inverse variation**, in **inverse proportion**, in **reciprocal proportion**) if each of the variables is directly proportional to the multiplicative inverse (reciprocal) of the other, or equivalently if their product is a constant. It follows that the variable y is inversely proportional to the variable x if there exists a non-zero constant k such that

$$y = \frac{k}{x}$$

The constant can be found by multiplying the original x variable and the original y variable.

As an example, the time taken for a journey is inversely proportional to the speed of travel; the time needed to dig a hole is (approximately) inversely proportional to the number of people digging.

The graph of two variables varying inversely on the Cartesian coordinate plane is a rectangular hyperbola. The product of the X and Y values of each point on the curve will equal the constant of proportionality (k). Since neither x nor y can equal zero (if k is non-zero), the graph will never cross either axis.

1.4 Hyperbolic coordinates

Main article: Hyperbolic coordinates

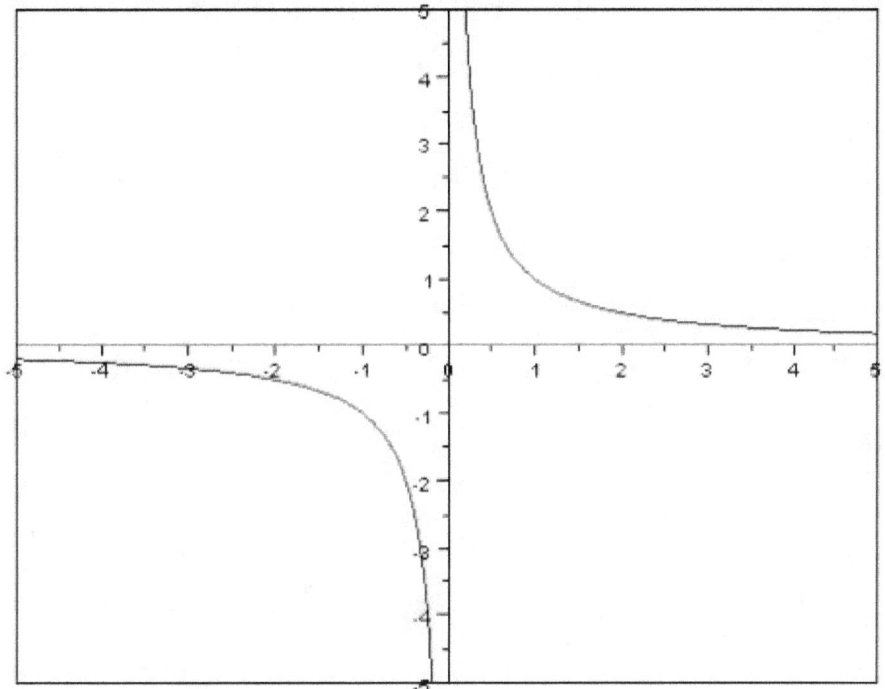

Inverse proportionality with a function of y=1/x.

The concepts of *direct* and *inverse* proportion lead to the location of points in the Cartesian plane by hyperbolic coordinates: the two coordinates correspond to the constant of direct proportionality that locates a point on a ray and the constant of inverse proportionality that locates a point on a hyperbola.

1.5   Exponential and logarithmic proportionality

A variable y is **exponentially proportional** to a variable x, if y is directly proportional to the exponential function of x, that is if there exist non-zero constants k and a such that

$$y = ka^x.$$

Likewise, a variable y is **logarithmically proportional** to a variable x, if y is directly proportional to the logarithm of x, that is if there exist non-zero constants k and a such that

$$y = k \log_a(x).$$

1.6 See also

- Correlation
- Eudoxus of Cnidus
- Golden ratio
- Proportional font
- Ratio
- Rule of three (mathematics)
- Sample size
- Similarity

1.6.1 Growth

- Linear growth
- Hyperbolic growth

Chapter 2

Function (mathematics)

In mathematics, a **function**[1] is a relation between a set of inputs and a set of permissible outputs with the property that each input is related to exactly one output. An example is the function that relates each real number x to its square x^2. The output of a function f corresponding to an input x is denoted by $f(x)$ (read "f of x"). In this example, if the input is -3, then the output is 9, and we may write $f(-3) = 9$. Likewise, if the input is 3, then the output is also 9, and we may write $f(3) = 9$. (The same output may be produced by more than one input, but each input gives only one output.) The input variable(s) are sometimes referred to as the argument(s) of the function.

Functions of various kinds are "the central objects of investigation"[2] in most fields of modern mathematics. There are many ways to describe or represent a function. Some functions may be defined by a formula or algorithm that tells how to compute the output for a given input. Others are given by a picture, called the graph of the function. In science, functions are sometimes defined by a table that gives the outputs for selected inputs. A function could be described implicitly, for example as the inverse to another function or as a solution of a differential equation.

The input and output of a function can be expressed as an ordered pair, ordered so that the first element is the input (or tuple of inputs, if the function takes more than one input), and the second is the output. In the example above, $f(x) = x^2$, we have the ordered pair $(-3, 9)$. If both input and output are real numbers, this ordered pair can be viewed as the Cartesian coordinates of a point on the graph of the function.

In modern mathematics,[3] a function is defined by its set of inputs, called the *domain*; a set containing the set of outputs, and possibly additional elements, as members, called its *codomain*; and the set of all input-output pairs, called its *graph*. Sometimes the codomain is called the function's "range", but more commonly the word "range" is used to mean, instead, specifically the set of outputs (this is also called the *image* of the function). For example, we could define a function using the rule $f(x) = x^2$ by saying that the domain and codomain are the real numbers, and that the graph consists of all pairs of real numbers (x, x^2). The image of this function is the set of non-negative real numbers. Collections of functions with the same domain and the same codomain are called function spaces, the properties of which are studied in such mathematical disciplines as real analysis, complex analysis, and functional analysis.

In analogy with arithmetic, it is possible to define addition, subtraction, multiplication, and division of functions, in those cases where the output is a number. Another important operation defined on functions is function composition, where the output from one function becomes the input to another function.

2.1   Introduction and examples

For an example of a function, let X be the set consisting of four shapes: a red triangle, a yellow rectangle, a green hexagon, and a red square; and let Y be the set consisting of five colors: red, blue, green, pink, and yellow. Linking each shape to its color is a function from X to Y: each shape is linked to a color (i.e., an element in Y), and each shape is "linked", or "mapped", to exactly one color. There is no shape that lacks a color and no shape that has two or more colors. This function will be referred to as the "color-of-the-shape function".

INPUT x

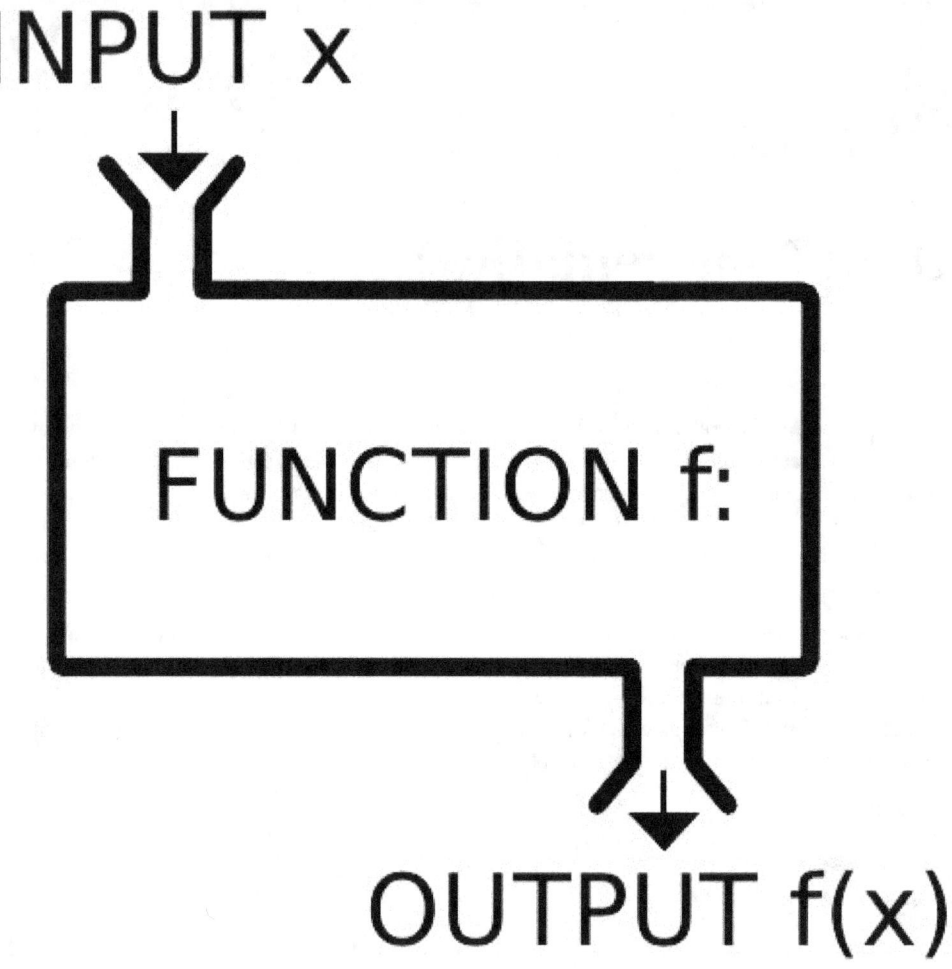

FUNCTION f:

OUTPUT f(x)

A function f takes an input x, and returns a single output f(x). One metaphor describes the function as a "machine" or "black box" that for each input returns a corresponding output.

The input to a function is called the argument and the output is called the value. The set of all permitted inputs to a given function is called the domain of the function, while the set of permissible outputs is called the codomain. Thus, the domain of the "color-of-the-shape function" is the set of the four shapes, and the codomain consists of the five colors. The concept of a function does *not* require that every possible output is the value of some argument, e.g. the color blue is not the color of any of the four shapes in X.

A second example of a function is the following: the domain is chosen to be the set of natural numbers (1, 2, 3, 4, ...), and the codomain is the set of integers (..., -3, -2, -1, 0, 1, 2, 3, ...). The function associates to any natural number n the number $4-n$. For example, to 1 it associates 3 and to 10 it associates -6.

A third example of a function has the set of polygons as domain and the set of natural numbers as codomain. The function associates a polygon with its number of vertices. For example, a triangle is associated with the number 3, a square with the number 4, and so on.

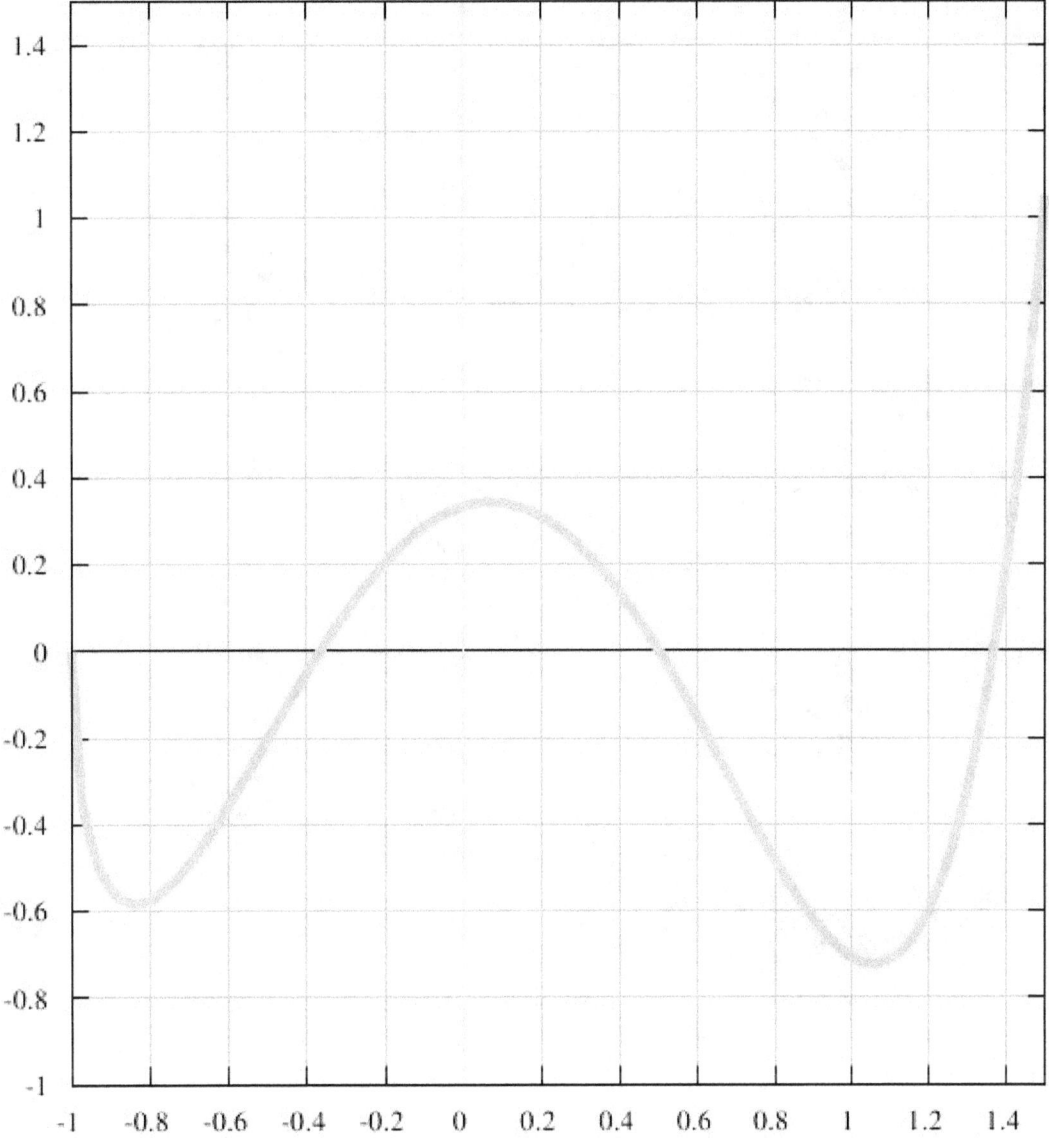

The red curve is the graph of a function f in the Cartesian plane, consisting of all points with coordinates of the form (x,f(x)). The property of having one output for each input is represented geometrically by the fact that each vertical line (such as the yellow line through the origin) has exactly one crossing point with the curve.

The term range is sometimes used either for the codomain or for the set of all the actual values a function has.

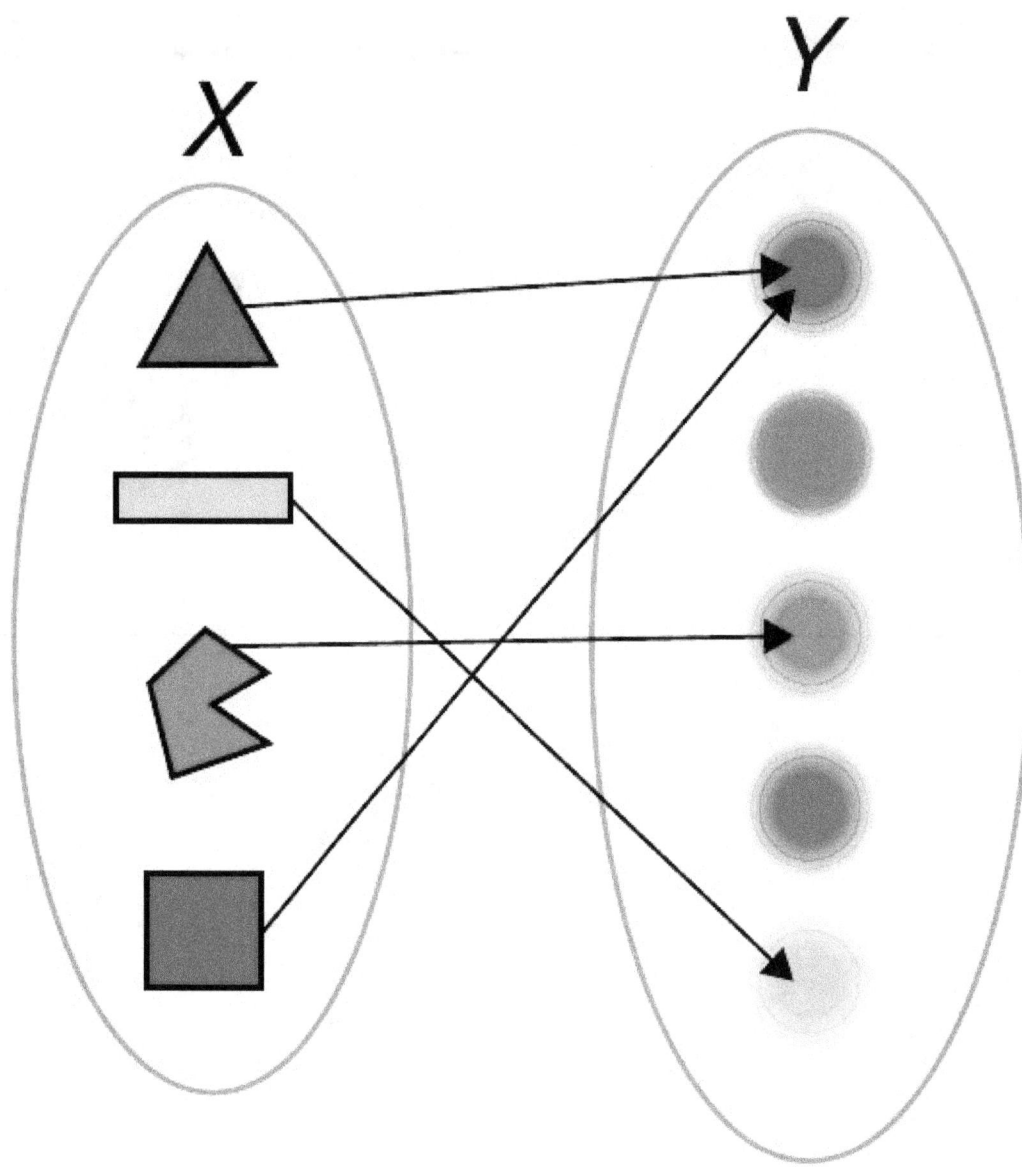

A function that associates to any of the four colored shapes its color.

2.2 Definition

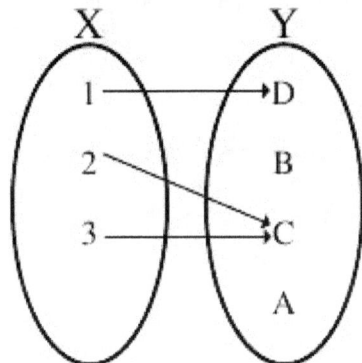

The above diagram represents a function with domain {1, 2, 3}, codomain {A, B, C, D} and set of ordered pairs {(1,D), (2,C), (3,C)}. The image is {C,D}.

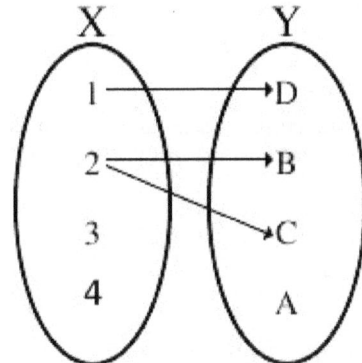

However, this second diagram does *not* represent a function. One reason is that 2 is the first element in more than one ordered pair. In particular, (2, B) and (2, C) are both elements of the set of ordered pairs. Another reason, sufficient by itself, is that 3 is not the first element (input) for any ordered pair. A third reason, likewise, is that 4 is not the first element of any ordered pair.

In order to avoid the use of the informally defined concepts of "rules" and "associates", the above intuitive explanation of functions is completed with a formal definition. This definition relies on the notion of the Cartesian product. The Cartesian product of two sets X and Y is the set of all ordered pairs, written (x, y), where x is an element of X and y is an element of Y. The x and the y are called the components of the ordered pair. The Cartesian product of X and Y is denoted by $X \times Y$.

A function f from X to Y is a subset of the Cartesian product $X \times Y$ subject to the following condition: every element of X is the first component of one and only one ordered pair in the subset.[4] In other words, for every x in X there is exactly one element y such that the ordered pair (x, y) is contained in the subset defining the function f. This formal definition is a precise rendition of the idea that to each x is associated an element y of Y, namely the uniquely specified element y with the property just mentioned.

Considering the "color-of-the-shape" function above, the set X is the domain consisting of the four shapes, while Y is the codomain consisting of five colors. There are twenty possible ordered pairs (four shapes times five colors), one of which is

    ("yellow rectangle", "red").

The "color-of-the-shape" function described above consists of the set of those ordered pairs,

(shape, color)

where the color is the actual color of the given shape. Thus, the pair ("red triangle", "red") is in the function, but the pair ("yellow rectangle", "red") is not.

2.3 Notation

For more details on this topic, see functional notation.

A function f with domain X and codomain Y is commonly denoted by

$$f: X \to Y$$

or

$$X \xrightarrow{f} Y.$$

In this context, the elements of X are called arguments of f. For each argument x, the corresponding unique y in the codomain is called the function value at x or the *image* of x under f. It is written as $f(x)$. One says that f associates y with x or maps x to y. This is abbreviated by

$$y = f(x).$$

A general function is often denoted by f. Special functions have names, for example, the signum function is denoted by sgn. Given a real number x, its image under the signum function is then written as $sgn(x)$. Here, the argument is denoted by the symbol x, but different symbols may be used in other contexts. For example, in physics, the velocity of some body, depending on the time, is denoted $v(t)$. The parentheses around the argument may be omitted when there is little chance of confusion, thus: $\sin x$; this is known as prefix notation.

In order to denote a specific function, the notation $\mapsto$ (an arrow with a bar at its tail) is used. For example, the above function reads

$$f: \mathbb{N} \to \mathbb{Z}$$
$$x \mapsto 4 - x$$

The first part can be read as:

- "f is a function from $\mathbb{N}$ (the set of natural numbers) to $\mathbb{Z}$ (the set of integers)" or
- "f is a $\mathbb{Z}$ -valued function of an $\mathbb{N}$ -valued variable".

The second part is read:

- "x maps to $4-x$."

In other words, this function has the natural numbers as domain, the integers as codomain. Strictly speaking, a function is properly defined only when the domain and codomain are specified. For example, the formula $f(x) = 4 - x$ alone (without specifying the codomain and domain) is not a properly defined function. Moreover, the function

$$g : \mathbb{Z} \to \mathbb{Z}$$
$$x \mapsto 4 - x.$$

(with different domain) is not considered the same function, even though the formulas defining f and g agree, and similarly with a different codomain. Despite that, many authors drop the specification of the domain and codomain, especially if these are clear from the context. So in this example many just write $f(x) = 4 - x$. Sometimes, the maximal possible domain is also understood implicitly: a formula such as $f(x) = \sqrt{x^2 - 5x + 6}$ may mean that the domain of f is the set of real numbers x where the square root is defined (in this case $x \leq 2$ or $x \geq 3$).[5]

To define a function, sometimes a dot notation is used in order to emphasize the functional nature of an expression without assigning a special symbol to the variable. For instance, $a(\cdot)^2$ stands for the function $x \mapsto ax^2$, $\int_a f(u)du$ stands for the integral function $x \mapsto \int_a^x f(u)du$, and so on.

2.4   Specifying a function

A function can be defined by any mathematical condition relating each argument (input value) to the corresponding output value. If the domain is finite, a function f may be defined by simply tabulating all the arguments x and their corresponding function values $f(x)$. More commonly, a function is defined by a formula, or (more generally) an algorithm — a recipe that tells how to compute the value of $f(x)$ given any x in the domain.

There are many other ways of defining functions. Examples include piecewise definitions, induction or recursion, algebraic or analytic closure, limits, analytic continuation, infinite series, and as solutions to integral and differential equations. The lambda calculus provides a powerful and flexible syntax for defining and combining functions of several variables. In advanced mathematics, some functions exist because of an axiom, such as the Axiom of Choice.

2.4.1   Graph

Main article: Graph of a function

The *graph* of a function is its set of ordered pairs F. This is an abstraction of the idea of a graph as a picture showing the function plotted on a pair of coordinate axes; for example, (3, 9), the point above 3 on the horizontal axis and to the right of 9 on the vertical axis, lies on the graph of $y=x^2$.

2.4.2   Formulas and algorithms

Different formulas or algorithms may describe the same function. For instance $f(x) = (x + 1)(x - 1)$ is exactly the same function as $f(x) = x^2 - 1$.[6] Furthermore, a function need not be described by a formula, expression, or algorithm, nor need it deal with numbers at all: the domain and codomain of a function may be arbitrary sets. One example of a function that acts on non-numeric inputs takes English words as inputs and returns the first letter of the input word as output.

As an example, the factorial function is defined on the nonnegative integers and produces a nonnegative integer. It is defined by the following inductive algorithm: 0! is defined to be 1, and $n!$ is defined to be $n(n - 1)!$ for all positive integers n. The factorial function is denoted with the exclamation mark (serving as the symbol of the function) after the variable (postfix notation).

2.4.3   Computability

Main article: computable function

Functions that send integers to integers, or finite strings to finite strings, can sometimes be defined by an algorithm, which gives a precise description of a set of steps for computing the output of the function from its input. Functions definable by an algorithm are called *computable functions*. For example, the Euclidean algorithm gives a precise process to compute the greatest common divisor of two positive integers. Many of the functions studied in the context of number theory are computable.

Fundamental results of computability theory show that there are functions that can be precisely defined but are not computable. Moreover, in the sense of cardinality, almost all functions from the integers to integers are not computable. The number of computable functions from integers to integers is countable, because the number of possible algorithms is. The number of all functions from integers to integers is higher: the same as the cardinality of the real numbers. Thus most functions from integers to integers are not computable. Specific examples of uncomputable functions are known, including the busy beaver function and functions related to the halting problem and other undecidable problems.

2.5 Basic properties

There are a number of general basic properties and notions. In this section, f is a function with domain X and codomain Y.

2.5.1 Image and preimage

Main article: Image (mathematics)

If A is any subset of the domain X, then $f(A)$ is the subset of the codomain Y consisting of all images of elements of

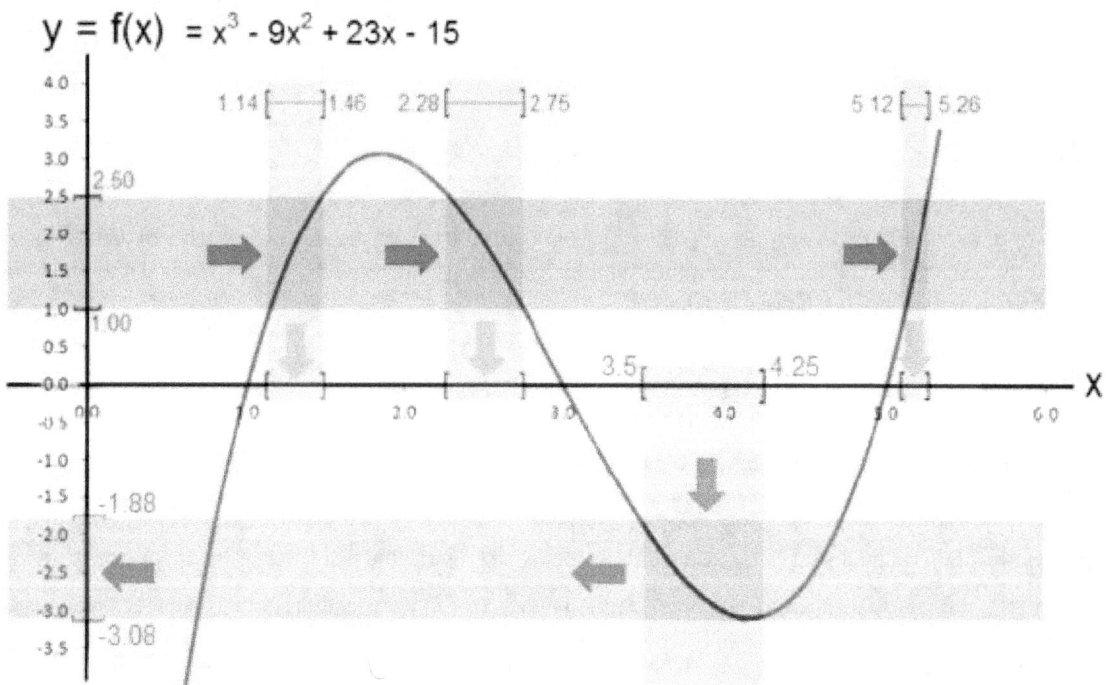

The graph of the function $f(x) = x^3 - 9x^2 + 23x - 15$. *The interval* $A = [3.5, 4.25]$ *is a subset of the domain, thus it is shown as part of the x-axis (green). The image of* A *is (approximately) the interval* $[-3.08, -1.88]$. *It is obtained by projecting to the y-axis (along the blue arrows) the intersection of the graph with the light green area consisting of all points whose x-coordinate is between 3.5 and 4.25. the part of the (vertical) y-axis shown in blue. The preimage of* $B = [1, 2.5]$ *consists of three intervals. They are obtained by projecting the intersection of the light red area with the graph to the x-axis.*

A. We say the $f(A)$ is the *image* of A under f. The *image* of f is given by $f(X)$. On the other hand, the *inverse image* (or *preimage, complete inverse image*) of a subset B of the codomain Y under a function f is the subset of the domain X defined by

$$f^{-1}(B) = \{x \in X : f(x) \in B\}.$$

So, for example, the preimage of $\{4, 9\}$ under the squaring function is the set $\{-3,-2,2,3\}$. The term range usually refers to the image,[7] but sometimes it refers to the codomain.

By definition of a function, the image of an element x of the domain is always a single element y of the codomain. Conversely, though, the preimage of a singleton set (a set with exactly one element) may in general contain any number of elements. For example, if $f(x) = 7$ (the constant function taking value 7), then the preimage of $\{5\}$ is the empty set but the preimage of $\{7\}$ is the entire domain. It is customary to write $f^{-1}(b)$ instead of $f^{-1}(\{b\})$, i.e.

$$f^{-1}(b) = \{x \in X : f(x) = b\}.$$

This set is sometimes called the fiber of b under f.

Use of $f(A)$ to denote the image of a subset $A \subseteq X$ is consistent so long as no subset of the domain is also an element of the domain. In some fields (e.g., in set theory, where ordinals are also sets of ordinals) it is convenient or even necessary to distinguish the two concepts; the customary notation is $f[A]$ for the set $\{ f(x): x \in A \}$. Likewise, some authors use square brackets to avoid confusion between the inverse image and the inverse function. Thus they would write $f^{-1}[B]$ and $f^{-1}[b]$ for the preimage of a set and a singleton.

2.5.2   Injective and surjective functions

A function is called *injective* (or *one-to-one*, or an injection) if $f(a) \neq f(b)$ for any two *different* elements a and b of the domain. It is called surjective (or *onto*) if $f(X) = Y$. That is, it is surjective if for every element y in the codomain there is an x in the domain such that $f(x) = y$. Finally f is called *bijective* if it is both injective and surjective. This nomenclature was introduced by the Bourbaki group.

The above "color-of-the-shape" function is not injective, since two distinct shapes (the red triangle and the red rectangle) are assigned the same value. Moreover, it is not surjective, since the image of the function contains only three, but not all five colors in the codomain.

2.5.3   Function composition

Main article: Function composition
 The *function composition* of two functions takes the output of one function as the input of a second one. More specifically, the composition of f with a function $g: Y \to Z$ is the function $g \circ f : X \to Z$ defined by

$$(g \circ f)(x) = g(f(x)).$$

That is, the value of x is obtained by first applying f to x to obtain $y = f(x)$ and then applying g to y to obtain $z = g(y)$. In the notation $g \circ f$, the function on the right, f, acts first and the function on the left, g acts second, reversing English reading order. The notation can be memorized by reading the notation as "g of f" or "g after f". The composition $g \circ f$ is only defined when the codomain of f is the domain of g. Assuming that, the composition in the opposite order $f \circ g$ need not be defined. Even if it is, i.e., if the codomain of f is the codomain of g, it is *not* in general true that

$$g \circ f = f \circ g.$$

That is, the order of the composition is important. For example, suppose $f(x) = x^2$ and $g(x) = x+1$. Then $g(f(x)) = x^2+1$, while $f(g(x)) = (x+1)^2$, which is x^2+2x+1, a different function.

2.5.4 Identity function

Main article: Identity function

The unique function over a set X that maps each element to itself is called the *identity function* for X, and typically denoted by id$X$. Each set has its own identity function, so the subscript cannot be omitted unless the set can be inferred from context. Under composition, an identity function is "neutral": if f is any function from X to Y, then

$$f \circ \text{id}_X = f,$$
$$\text{id}_Y \circ f = f.$$

2.5.5 Restrictions and extensions

Main article: Restriction (mathematics)

Informally, a *restriction* of a function f is the result of trimming its domain. More precisely, if S is any subset of X, the restriction of f to S is the function $f|S$ from S to Y such that $f|S(s) = f(s)$ for all s in S. If g is a restriction of f, then it is said that f is an *extension* of g.

The *overriding* of $f: X \rightarrow Y$ by $g: W \rightarrow Y$ (also called *overriding union*) is an extension of g denoted as $(f \oplus g): (X \cup W) \rightarrow Y$. Its graph is the set-theoretical union of the graphs of g and $f|X \setminus W$. Thus, it relates any element of the domain of g to its image under g, and any other element of the domain of f to its image under f. Overriding is an associative operation; it has the empty function as an identity element. If $f|X \cap W$ and $g|X \cap W$ are pointwise equal (e.g., the domains of f and g are disjoint), then the union of f and g is defined and is equal to their overriding union. This definition agrees with the definition of union for binary relations.

2.5.6 Inverse function

Main article: Inverse function

An *inverse function* for f, denoted by f^{-1}, is a function in the opposite direction, from Y to X, satisfying

$$f \circ f^{-1} = \text{id}_Y, f^{-1} \circ f = \text{id}_X.$$

That is, the two possible compositions of f and f^{-1} need to be the respective identity maps of X and Y.

As a simple example, if f converts a temperature in degrees Celsius C to degrees Fahrenheit F, the function converting degrees Fahrenheit to degrees Celsius would be a suitable f^{-1}.

$$f(C) = \frac{9}{5}C + 32$$
$$f^{-1}(F) = \frac{5}{9}(F - 32)$$

Such an inverse function exists if and only if f is bijective. In this case, f is called invertible. The notation $g \circ f$ (or, in some texts, just gf) and f^{-1} are akin to multiplication and reciprocal notation. With this analogy, identity functions are like the multiplicative identity, 1, and inverse functions are like reciprocals (hence the notation).

2.6   Types of functions

For a more extensive list, see list of types of functions.

2.6.1   Real-valued functions

A real-valued function f is one whose codomain is the set of real numbers or a subset thereof. If, in addition, the domain is also a subset of the reals, f is a real valued function of a real variable. The study of such functions is called real analysis.

Real-valued functions enjoy so-called pointwise operations. That is, given two functions

$$f, g: X \to Y$$

where Y is a subset of the reals (and X is an arbitrary set), their (pointwise) sum $f+g$ and product $f \cdot g$ are functions with the same domain and codomain. They are defined by the formulas:

$$(f + g)(x) = f(x) + g(x).$$
$$(f \cdot g)(x) = f(x) \cdot g(x).$$

In a similar vein, complex analysis studies functions whose domain and codomain are both the set of complex numbers. In most situations, the domain and codomain are understood from context, and only the relationship between the input and output is given, but if $f(x) = \sqrt{x}$, then in real variables the domain is limited to non-negative numbers.

The following table contains a few particularly important types of real-valued functions:

2.6.2   Further types of functions

Further information: List of mathematical functions

There are many other special classes of functions that are important to particular branches of mathematics, or particular applications. Here is a partial list:

- differentiable, integrable

- polynomial, rational

- algebraic, transcendental

- odd or even

- convex, monotonic

- holomorphic, meromorphic, entire

- vector-valued

- computable

2.7 Function spaces

Main article: Function space

The set of all functions from a set X to a set Y is denoted by $X \to Y$, by $[X \to Y]$, or by Y^X. The latter notation is motivated by the fact that, when X and Y are finite and of size $|X|$ and $|Y|$, then the number of functions $X \to Y$ is $|Y^X| = |Y|^{|X|}$. This is an example of the convention from enumerative combinatorics that provides notations for sets based on their cardinalities. If X is infinite and there is more than one element in Y then there are uncountably many functions from X to Y, though only countably many of them can be expressed with a formula or algorithm.

2.7.1 Currying

Main article: Currying

An alternative approach to handling functions with multiple arguments is to transform them into a chain of functions that each takes a single argument. For instance, one can interpret Add(3,5) to mean "first produce a function that adds 3 to its argument, and then apply the 'Add 3' function to 5". This transformation is called currying: Add 3 is curry(Add) applied to 3. There is a bijection between the function spaces $C^{A \times B}$ and $(C^B)^A$.

When working with curried functions it is customary to use prefix notation with function application considered left-associative, since juxtaposition of multiple arguments—as in $(f\ x\ y)$—naturally maps to evaluation of a curried function. Conversely, the $\to$ and $\mapsto$ symbols are considered to be right-associative, so that curried functions may be defined by a notation such as $f: \mathbf{Z} \to \mathbf{Z} \to \mathbf{Z} = x \mapsto y \mapsto x \cdot y$.

2.8 Variants and generalizations

2.8.1 Alternative definition of a function

The above definition of "a function from X to Y" is generally agreed on, however there are two different ways a "function" is normally defined where the domain X and codomain Y are not explicitly or implicitly specified. Usually this is not a problem as the domain and codomain normally will be known. With one definition saying the function defined by $f(x) = x^2$ on the reals does not completely specify a function as the codomain is not specified, and in the other it is a valid definition.

In the other definition a function is defined as a set of ordered pairs where each first element only occurs once. The domain is the set of all the first elements of a pair and there is no explicit codomain separate from the image.[8][9] Concepts like surjective have to be refined for such functions, more specifically by saying that a (given) function is *surjective on a (given) set* if its image equals that set. For example, we might say a function f is surjective on the set of real numbers.

If a function is defined as a set of ordered pairs with no specific codomain, then $f: X \to Y$ indicates that f is a function whose domain is X and whose image is a subset of Y. This is the case in the ISO standard.[7] Y may be referred to as the codomain but then any set including the image of f is a valid codomain of f. This is also referred to by saying that "f maps X into Y".[7] In some usages X and Y may subset the ordered pairs, e.g. the function f on the real numbers such that $y=x^2$ when used as in $f: [0,4] \to [0,4]$ means the function defined only on the interval $[0,2]$.[10] With the definition of a function as an ordered triple this would always be considered a partial function.

An alternative definition of the composite function $g(f(x))$ defines it for the set of all x in the domain of f such that $f(x)$ is in the domain of g.[11] Thus the real square root of $-x^2$ is a function only defined at 0 where it has the value 0.

Functions are commonly defined as a type of relation. A relation from X to Y is a set of ordered pairs (x, y) with $x \in X$ and $y \in Y$. A function from X to Y can be described as a relation from X to Y that is left-total and right-unique. However when X and Y are not specified there is a disagreement about the definition of a relation that parallels that for functions. Normally a relation is just defined as a set of ordered pairs and a correspondence is defined as a triple (X, Y, F), however the distinction between the two is often blurred or a relation is never referred to without specifying the two sets. The definition of a function as a triple defines a function as a type of correspondence, whereas the definition of a function as a set of ordered pairs defines a function as a type of relation.

Many operations in set theory, such as the power set, have the class of all sets as their domain, and therefore, although they are informally described as functions, they do not fit the set-theoretical definition outlined above, because a class is not necessarily a set. However some definitions of relations and functions define them as classes of pairs rather than sets of pairs and therefore do include the power set as a function.[12]

2.8.2   Partial and multi-valued functions

In some parts of mathematics, including recursion theory and functional analysis, it is convenient to study *partial functions* in which some values of the domain have no association in the graph; i.e., single-valued relations. For example, the function f such that $f(x) = 1/x$ does not define a value for $x = 0$, since division by zero is not defined. Hence f is only a partial function from the real line to the real line. The term total function can be used to stress the fact that every element of the domain does appear as the first element of an ordered pair in the graph. In other parts of mathematics, non-single-valued relations are similarly conflated with functions: these are called *multivalued functions*, with the corresponding term single-valued function for ordinary functions.

2.8.3   Functions with multiple inputs and outputs

The concept of function can be extended to an object that takes a combination of two (or more) argument values to a single result. This intuitive concept is formalized by a function whose domain is the Cartesian product of two or more sets.

For example, consider the function that associates two integers to their product: $f(x, y) = x \cdot y$. This function can be defined formally as having domain $\mathbf{Z} \times \mathbf{Z}$, the set of all integer pairs; codomain $\mathbf{Z}$; and, for graph, the set of all pairs $((x,y), x \cdot y)$. Note that the first component of any such pair is itself a pair (of integers), while the second component is a single integer.

The function value of the pair (x,y) is $f((x,y))$. However, it is customary to drop one set of parentheses and consider $f(x,y)$ a function of two variables, x and y. Functions of two variables may be plotted on the three-dimensional Cartesian as ordered triples of the form $(x,y,f(x,y))$.

The concept can still further be extended by considering a function that also produces output that is expressed as several variables. For example, consider the integer divide function, with domain $\mathbf{Z} \times \mathbf{N}$ and codomain $\mathbf{Z} \times \mathbf{N}$. The resultant (quotient, remainder) pair is a single value in the codomain seen as a Cartesian product.

Binary operations

The familiar binary operations of arithmetic, addition and multiplication, can be viewed as functions from $\mathbf{R} \times \mathbf{R}$ to $\mathbf{R}$. This view is generalized in abstract algebra, where n-ary functions are used to model the operations of arbitrary algebraic structures. For example, an abstract group is defined as a set X and a function f from $X \times X$ to X that satisfies certain properties.

Traditionally, addition and multiplication are written in the infix notation: $x+y$ and $x \times y$ instead of $+(x, y)$ and $\times(x, y)$.

2.8.4   Functors

The idea of structure-preserving functions, or homomorphisms, led to the abstract notion of morphism, the key concept of category theory. In fact, functions $f: X \rightarrow Y$ are the morphisms in the category of sets, including the empty set: if the domain X is the empty set, then the subset of $X \times Y$ describing the function is necessarily empty, too. However, this is still a well-defined function. Such a function is called an empty function. In particular, the identity function of the empty set is defined, a requirement for sets to form a category.

The concept of categorification is an attempt to replace set-theoretic notions by category-theoretic ones. In particular, according to this idea, sets are replaced by categories, while functions between sets are replaced by functors.[13]

2.9   History

Main article: History of the function concept

2.10   See also

- Associative array
- Functional
- Functional decomposition
- Function fitting
- Functional predicate
- Functional programming
- Generalized function
- Implicit function
- List of functions
- Multivalued function
- Parametric equation

2.11   Notes

[1] The words **map** or **mapping**, **transformation**, **correspondence**, and **operator** are often used synonymously. Halmos 1970, p. 30.

[2] Spivak 2008, p. 39.

[3] MacLane, Saunders, Birkhoff, Garrett (1967). *Algebra* (First ed.). New York. Macmillan. pp. 1–13.

[4] Hamilton, A. G. *Numbers, sets, and axioms: the apparatus of mathematics*. Cambridge University Press. p. 83. ISBN 0-521-24509-5.

[5] Bloch 2011, p. 133.

[6] Hartley Rogers, Jr (1987). *Theory of Recursive Functions and Effective Computation*. MIT Press. pp. 1–2. ISBN 0-262-68052-1.

[7] *Quantities and Units - Part 2: Mathematical signs and symbols to be used in the natural sciences and technology*, page 15. ISO 80000-2 (ISO/IEC 2009-12-01)

[8] Apostol, Tom (1967). *Calculus vol 1*. John Wiley. p. 53. ISBN 0-471-00005-1.

[9] Heins, Maurice (1968). *Complex function theory*. Academic Press. p. 4.

[10] Bartle 1967, p. 13.

[11] Bartle 1967, p. 21.

[12] Tarski, Alfred; Givant, Steven (1987). *A formalization of set theory without variables*. American Mathematical Society. p. 3. ISBN 0-8218-1041-3.

[13] John C. Baez; James Dolan (1998). "Categorification". arXiv:math/9802029.

2.12 References

- Bartle, Robert (1967). *The Elements of Real Analysis*. John Wiley & Sons.

- Bloch, Ethan D. (2011). *Proofs and Fundamentals: A First Course in Abstract Mathematics*. Springer. ISBN 978-1-4419-7126-5.

- Halmos, Paul R. (1970). *Naive Set Theory*. Springer-Verlag. ISBN 0-387-90092-6.

- Spivak, Michael (2008). *Calculus* (4th ed.). Publish or Perish. ISBN 978-0-914098-91-1.

2.13 Further reading

- Anton, Howard (1980). *Calculus with Analytical Geometry*. Wiley. ISBN 978-0-471-03248-9.

- Bartle, Robert G. (1976). *The Elements of Real Analysis* (2nd ed.). Wiley. ISBN 978-0-471-05464-1.

- Dubinsky, Ed; Harel, Guershon (1992). *The Concept of Function: Aspects of Epistemology and Pedagogy*. Mathematical Association of America. ISBN 0-88385-081-8.

- Hammack, Richard (2009). "12. Functions" (PDF). *Book of Proof*. Virginia Commonwealth University. Retrieved 2012-08-01.

- Husch, Lawrence S. (2001). *Visual Calculus*. University of Tennessee. Retrieved 2007-09-27.

- Katz, Robert (1964). *Axiomatic Analysis*. D. C. Heath and Company.

- Kleiner, Israel (1989). *Evolution of the Function Concept: A Brief Survey. The College Mathematics Journal* **20** (4) (Mathematical Association of America). pp. 282–300. doi:10.2307/2686848. JSTOR 2686848.

- Lützen, Jesper (2003). "Between rigor and applications: Developments in the concept of function in mathematical analysis". In Roy Porter, ed. *The Cambridge History of Science: The modern physical and mathematical sciences*. Cambridge University Press. ISBN 0521571995. An approachable and diverting historical presentation.

- Malik, M. A. (1980). *Historical and pedagogical aspects of the definition of function. International Journal of Mathematical Education in Science and Technology* **11** (4). pp. 489–492. doi:10.1080/0020739800110404.

- Reichenbach, Hans (1947) *Elements of Symbolic Logic*, Dover Publishing Inc., New York NY, ISBN 0-486-24004-5.

- Ruthing, D. (1984). *Some definitions of the concept of function from Bernoulli, Joh. to Bourbaki, N. Mathematical Intelligencer* **6** (4). pp. 72–77.

- Thomas, George B.; Finney, Ross L. (1995). *Calculus and Analytic Geometry* (9th ed.). Addison-Wesley. ISBN 978-0-201-53174-9.

2.14 External links

- Khan Academy: Functions, free online micro lectures

- Hazewinkel, Michiel, ed. (2001), "Function", *Encyclopedia of Mathematics*, Springer, ISBN 978-1-55608-010-4

- Weisstein, Eric W., "Function", *MathWorld*.

- The Wolfram Functions Site gives formulae and visualizations of many mathematical functions.

- Shodor: Function Flyer, interactive Java applet for graphing and exploring functions.

- xFunctions, a Java applet for exploring functions graphically.

- Draw Function Graphs, online drawing program for mathematical functions.

- Functions from cut-the-knot.

- Function at ProvenMath.

- Comprehensive web-based function graphing & evaluation tool.

- Abstractmath.org articles on functions

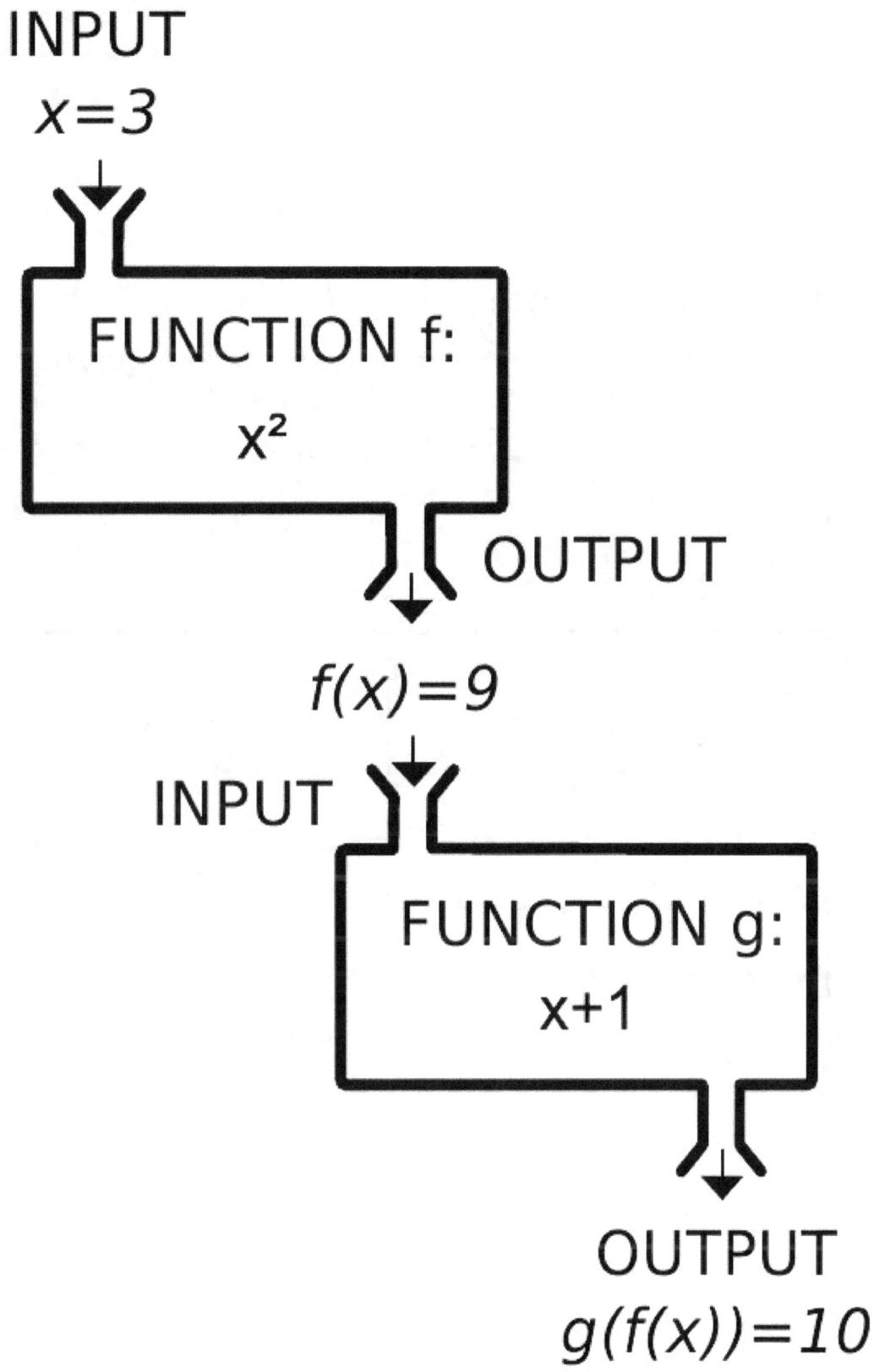

INPUT

x=3

FUNCTION f:

x²

OUTPUT

f(x)=9

INPUT

FUNCTION g:

x+1

OUTPUT

g(f(x))=10

A composite function g(f(x)) can be visualized as the combination of two "machines". The first takes input x and outputs f(x). The second takes f(x) and outputs g(f(x)).

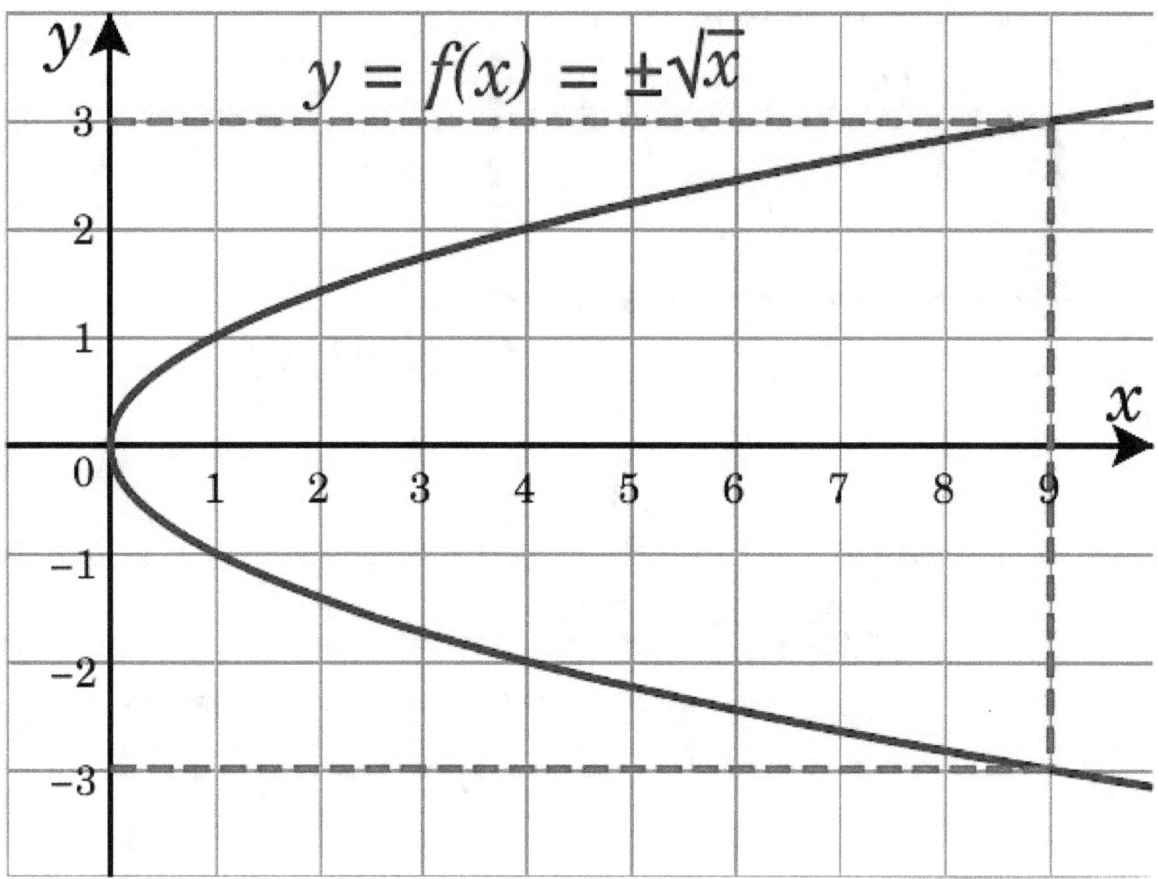

$f(x) = \pm\sqrt{x}$ *is not a function in the proper sense, but a multi-valued function: it assigns to each positive real number* x *two values: the (positive) square root of* x, *and* $-\sqrt{x}$.

Chapter 3

Product (mathematics)

In mathematics, a **product** is the result of multiplying, or an expression that identifies factors to be multiplied. Thus, for instance, 6 is the product of 2 and 3 (the result of multiplication), and $x \cdot (2 + x)$ is the product of x and $(2 + x)$ (indicating that the two factors should be multiplied together).

The order in which real or complex numbers are multiplied has no bearing on the product; this is known as the commutative law of multiplication. When matrices or members of various other associative algebras are multiplied, the product usually depends on the order of the factors. Matrix multiplication, for example, and multiplication in other algebras is in general non-commutative.

There are many different kinds of products in mathematics: besides being able to multiply just numbers, polynomials or matricies, one can also define products on many different algebraic structures. An overview of these different kinds of products is given here.

3.1 Product of two numbers

Main article: multiplication

3.1.1 Product of two natural numbers

Placing several stones into a rectangular pattern with r rows and s columns gives

$$r \cdot s = \sum_{i=1}^{s} r = \sum_{j=1}^{r} s$$

stones.

3.1.2 Product of two integers

Integers allow positive and negative numbers. The two numbers are multiplied just like natural numbers, except we need an additional rule for the signs:

$\cdot$	$-$	$+$
$-$	$+$	$-$
$+$	$-$	$+$

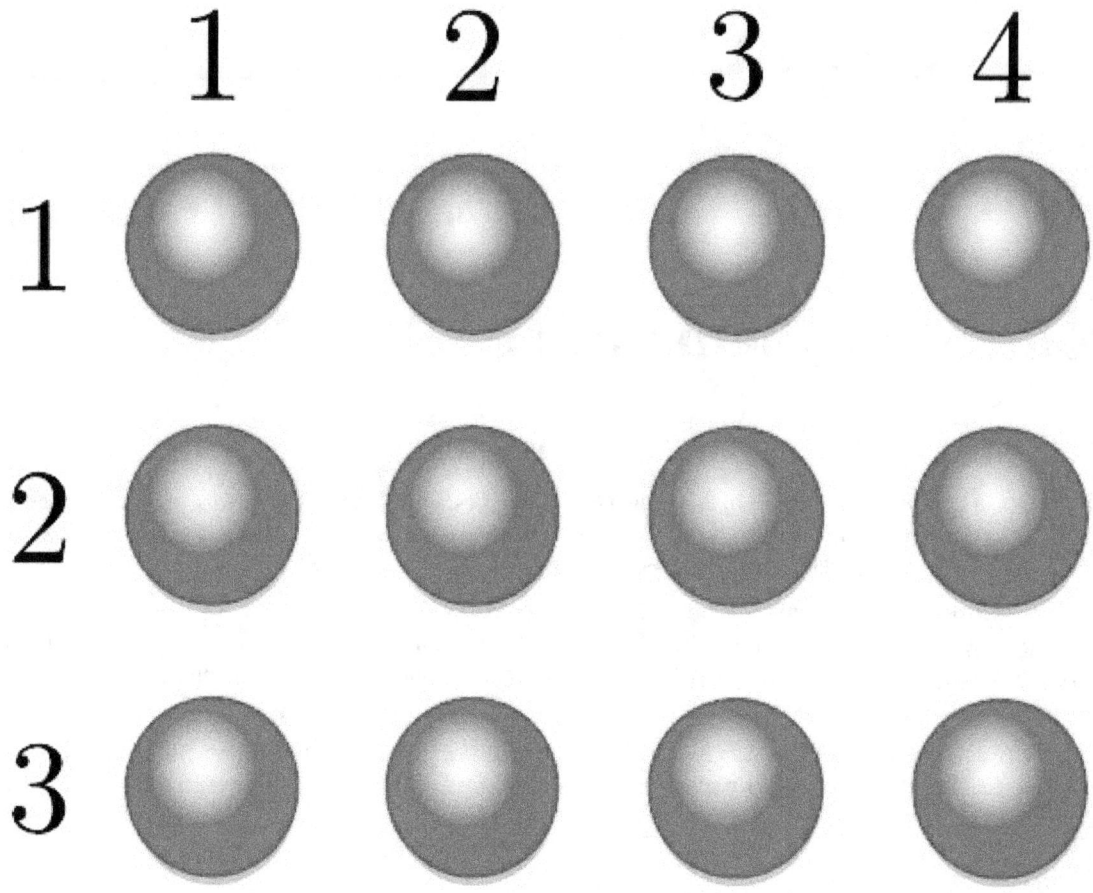

3 by 4 is 12

In words, we have:

- Minus times Minus gives Plus
- Minus times Plus gives Minus
- Plus times Minus gives Minus
- Plus times Plus gives Plus

3.1.3   Product of two fractions

Two fractions can be multiplied by multiplying their numerators and denominators:

$$\frac{z}{n} \cdot \frac{z'}{n'} = \frac{z \cdot z'}{n \cdot n'}$$

3.1.4   Product of two real numbers

For a rigorous definition of the product of two real numbers see Construction of the real numbers.

3.1.5   Product of two complex numbers

Two complex numbers can be multiplied by the distributive law and the fact that $i^2 = -1$, as follows:

$$(a + bi) \cdot (c + di) = a \cdot c + a \cdot di + b \cdot ci + b \cdot d \cdot i^2$$
$$= (a \cdot c - b \cdot d) + (a \cdot d + b \cdot c)i$$

Geometric meaning of complex multiplication

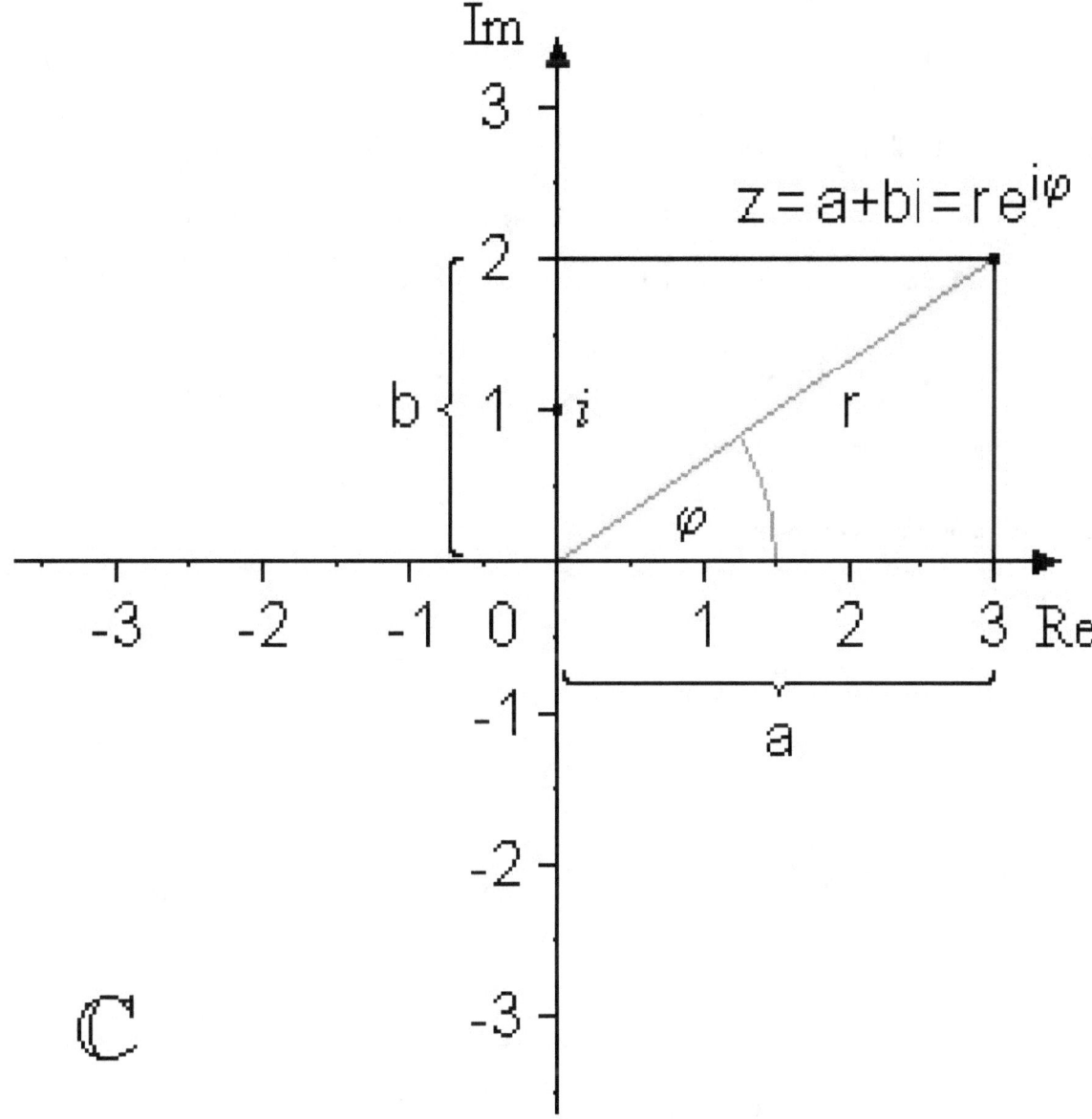

A complex number in polar coordinates.

Complex numbers can be written in polar coordinates:

$$a + b\,\mathrm{i} = r \cdot (\cos(\varphi) + \mathrm{i}\sin(\varphi)) = r \cdot \mathrm{e}^{\mathrm{i}\varphi}$$

Furthermore,

$$c + d\,\mathrm{i} = s \cdot (\cos(\psi) + \mathrm{i}\sin(\psi)) = s \cdot \mathrm{e}^{\mathrm{i}\psi}$$

$$(a \cdot c - b \cdot d) + (a \cdot d + b \cdot c)\,\mathrm{i} = r \cdot s \cdot (\cos(\varphi + \psi) + \mathrm{i}\sin(\varphi + \psi)) = r \cdot s \cdot \mathrm{e}^{\mathrm{i}(\varphi + \psi)}$$

The geometric meaning is that we multiply the magnitudes and add the angles.

3.1.6 Product of two quaternions

The product of two quaternions can be found in the article on quaternions. However, it is interesting to note that in this case, $a \cdot b$ and $b \cdot a$ are different.

3.2 Product of sequences

The product operator for the product of a sequence is denoted by the capital Greek letter Pi $\prod$ (in analogy to the use of the capital Sigma $\sum$ as summation symbol). The product of a sequence consisting of only one number is just that number itself. The product of no factors at all is known as the empty product, and is equal to 1.

3.3 Commutative rings

Commutative rings have a product operation.

3.3.1 Residue classes of integers

Main article: residue class

Residue classes in the rings $\mathbb{Z}/N\mathbb{Z}$ can be added:

$$(a + N\mathbb{Z}) + (b + N\mathbb{Z}) = a + b + N\mathbb{Z}$$

and multiplied:

$$(a + N\mathbb{Z}) \cdot (b + N\mathbb{Z}) = a \cdot b + N\mathbb{Z}$$

3.3.2 Rings of functions

Main article: ring of functions

Functions to the real numbers can be added or multiplied by adding or multiplying their outputs:

$$(f + g)(m) := f(m) + g(m)$$
$$(f \cdot g)(m) := f(m) \cdot g(m)$$

3.3.3   Convolution

Main article: convolution

Two functions from the reals to itself can be multiplied in another way, called the convolution.

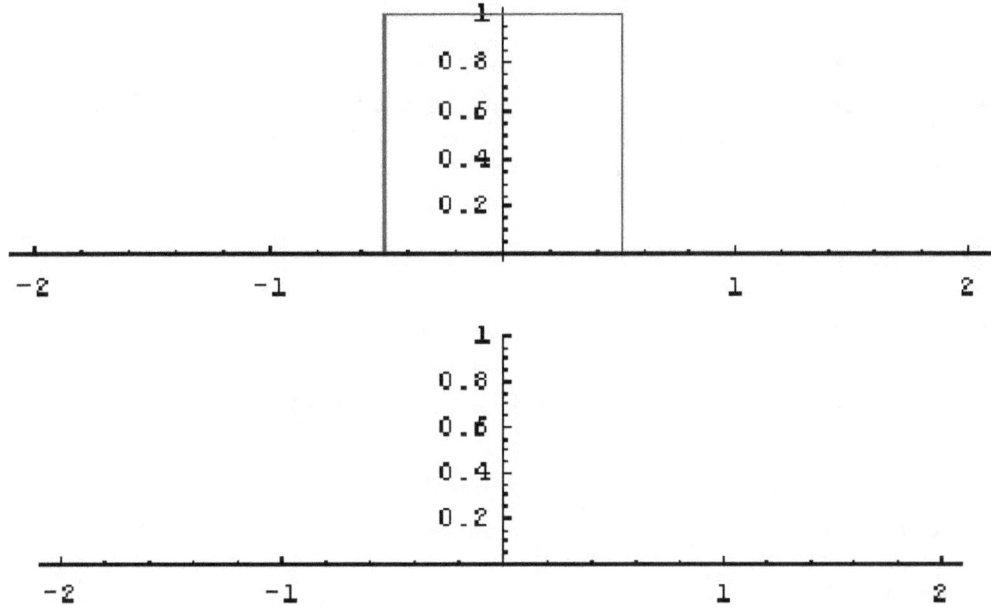

The convolution of the square wave with itself gives the triangular function

If : $\int\limits_{-\infty}^{\infty} |f(t)|\,\mathrm{d}t < \infty$ and $\int\limits_{-\infty}^{\infty} |g(t)|\,\mathrm{d}t < \infty$

then the integral

$$(f * g)(t) := \int\limits_{-\infty}^{\infty} f(\tau) \cdot g(t - \tau)\,\mathrm{d}\tau$$

is well defined and is called the convolution.

Under the Fourier transform, convolution becomes point-wise function multiplication.

3.3.4   Polynomial rings

Main article: polynomial ring

The product of two polynomials is given by the following:

$$\left(\sum_{i=0}^{n} a_i X^i\right) \cdot \left(\sum_{j=0}^{m} b_j X^j\right) = \sum_{k=0}^{n+m} c_k X^k$$

with

$$c_k = \sum_{i+j=k} a_i \cdot b_j$$

3.4 Products in linear algebra

There are many different kinds of products in linear algebra: some of these have confusingly similar names (outer product, exterior product) but have very different meanings. Others have very different names (outer product, tensor product, Kronecker product) but convey essentially the same idea. A brief overview of these is given here.

3.4.1 Scalar multiplication

Main article: scalar multiplication

By the very definition of a vector space, one can form the product of any scalar with any vector, giving a map $\mathbb{R} \times V \to V$.

3.4.2 Scalar product

Main article: scalar product

A scalar product is a bilinear map:

$$\cdot : V \times V \to \mathbb{R}$$

with the following conditions, that $v \cdot v > 0$ for all $0 \neq v \in V$.

From the scalar product, one can define a norm by letting $\|v\| := \sqrt{v \cdot v}$.

The scalar product also allows one to define an angle between two vectors:

$$\cos \angle(v, w) = \frac{v \cdot w}{\|v\| \cdot \|w\|}$$

In n-dimensional Euclidean space, the standard scalar product (called the dot product) is given by:

$$\left(\sum_{i=1}^{n} \alpha_i e_i \right) \cdot \left(\sum_{i=1}^{n} \beta_i e_i \right) = \sum_{i=1}^{n} \alpha_i \beta_i$$

3.4.3 Cross product in 3-dimensional space

Main article: cross product

The cross product of two vectors in 3-dimensions is a vector perpendicular to the two factors, with length equal to the area of the parallelogram spanned by the two factors.

The cross product can also be expressed as the formal[lower-alpha 1] determinant:

$$\mathbf{u} \times \mathbf{v} = \begin{vmatrix} \mathbf{i} & \mathbf{j} & \mathbf{k} \\ u_1 & u_2 & u_3 \\ v_1 & v_2 & v_3 \end{vmatrix}$$

3.4.4   Composition of linear mappings

Main article: function composition

A linear mapping can be defined as a function f between two vector spaces V and W with underlying field $\mathbf{F}$, satisfying[1]

$$f(t_1 x_1 + t_2 x_2) = t_1 f(x_1) + t_2 f(x_2), \forall x_1, x_2 \in V, \forall t_1, t_2 \in \mathbb{F}.$$

If one only considers finite dimensional vector spaces, then

$$f(\mathbf{v}) = f(v_i \mathbf{b_V}^i) = v_i f(\mathbf{b_V}^i) = f^i{}_j v_i \mathbf{b_W}^j.$$

in which $\mathbf{b_V}$ and $\mathbf{b_W}$ denote the bases of V and W, and v_i denotes the component of $\mathbf{v}$ on $\mathbf{b_V}^i$, and Einstein summation convention is applied.

Now we consider the composition of two linear mappings between finite dimensional vector spaces. Let the linear mapping f map V to W, and let the linear mapping g map W to U. Then one can get

$$g \circ f(\mathbf{v}) = g(f^i{}_j v_i \mathbf{b_W}^j) = g^j{}_k f^i{}_j v_i \mathbf{b_U}^k.$$

Or in matrix form:

$$g \circ f(\mathbf{v}) = \mathbf{GFv}.$$

in which the i-row, j-column element of $\mathbf{F}$, denoted by Fij, is $f^i i$, and $Gij=g^j i$.

The composition of more than two linear mappings can be similarly represented by a chain of matrix multiplication.

3.4.5   Product of two matrices

Main article: matrix product

Given two matrices

$$A = (a_{i,j})_{i=1\ldots s; j=1\ldots r} \in \mathbb{R}^{s \times r} \text{ and } B = (b_{j,k})_{j=1\ldots r; k=1\ldots t} \in \mathbb{R}^{r \times t}$$

their product is given by

$$B \cdot A = \left(\sum_{j=1}^{r} a_{i,j} \cdot b_{j,k} \right)_{i=1\ldots s; k=1\ldots t} \in \mathbb{R}^{s \times t}$$

3.4.6 Composition of linear functions as matrix product

There is a relationship between the composition of linear functions and the product of two matrices. To see this, let r = dim(U), s = dim(V) and t = dim(W) be the (finite) dimensions of vector spaces U, V und W. Let $\mathcal{U} = \{u_1, \ldots u_r\}$ be a basis von U. $\mathcal{V} = \{v_1, \ldots v_s\}$ be a basis of V und $\mathcal{W} = \{w_1, \ldots w_t\}$ be a basis of W. In terms of this basis, let $A = M_V^{\mathcal{U}}(f) \in \mathbb{R}^{s \times r}$ be the matrix representing f : U → V and $B = M_W^{\mathcal{V}}(g) \in \mathbb{R}^{r \times t}$ be the matrix representing g : V → W. Then

$$B \cdot A = M_W^{\mathcal{U}}(g \circ f) \in \mathbb{R}^{s \times t}$$

is the matrix representing $g \circ f : U \to W$.

In other words: the matrix product is the description in coordinates of the composition of linear functions.

3.4.7 Tensor product of vector spaces

Main article: Tensor product

Given two finite dimensional vector spaces *V* and *W*, the tensor product of them can be defined as a (2,0)-tensor satisfying:

$$V \otimes W(v, m) = V(v)W(w), \forall v \in V^*, \forall w \in W^*,$$

where V^* and W^* denote the dual spaces of *V* and *W*.[2]

For inifinite-dimensional vector spaces, one also has the:

- Tensor product of Hilbert spaces
- Topological tensor product.

The tensor product, outer product and Kronecker product all convey the same general idea. The differences between these are that the Kronecker product is just a tensor product of matrices, with respect to a previously-fixed basis, whereas the tensor product is usually given in it's intrinsic definition. The outer product is simply the Kronecker product, limited to vectors (instead of matrices).

3.4.8 The class of all objects with a tensor product

In general, whenever one has two mathematical objects that can be combined in a way that behaves like a linear algebra tensor product, then this can be most generally understood as the internal product of a monoidal category. That is, the monoidal category captures precisely the meaning of a tensor product; it captures exactly the notion of why it is that tensor products behave the way they do. More precisely, a monoidal category is the class of all things (of a given type) that have a tensor product.

3.4.9 Other products in linear algebra

Other kinds of products in linear algebra include:

- Hadamard product
- Kronecker product

- The product of tensors:

  - Wedge product or exterior product
  - Interior product
  - Outer product
  - Tensor product

3.5   Cartesian product

In set theory, a Cartesian product is a mathematical operation which returns a set (or **product set**) from multiple sets. That is, for sets A and B, the Cartesian product $A \times B$ is the set of all ordered pairs (a, b) where a $\in A$ and b $\in B$.[3]

The class of all things (of a given type) that have Cartesian products is called a Cartesian cateogry. Many of these are Cartesian closed categories. Sets are an example of such objects.

3.6   Empty product

The empty product on numbers and most algebraic structures has the value of 1 (the identity element of multiplication) just like the empty sum has the value of 0 (the identity element of addition). However, the concept of the empty product is more general, and requires special treatment in logic, set theory, computer programming and category theory.

3.7   Products over other algebraic structures

Products over other kinds of algebraic structures include:

- the Cartesian product of sets,

- the product of groups, and also the semidirect product, knit product and wreath product,

- the free product of groups

- the product of rings,

- the product of ideals,

- the product of topological spaces,

- the Wick product of random variables,

- the cap, cup and slant product in algebraic topology,

- the smash product and wedge sum (sometimes called the wedge product) in homotopy.

A few of the above products are examples of the general notion of an internal product in a monoidal category; the rest are describable by the general notion of a product in category theory.

3.8 Products in category theory

All of the previous examples are special cases or examples of the general notion of a product. For the general treatment of the concept of a product, see product (category theory), which describes how to combine two objects of some kind to create an object, possibly of a different kind. But also, in category theory, one has:

- the fiber product or pullback,
- the product category, a category that is the product of categories.
- the ultraproduct, in model theory.
- the internal product of a monoidal category, which captures the essence of a tensor product.

3.9 Other products

- A function's product integral (as a continuous equivalent to the product of a sequence or the multiplicative version of the (normal/standard/additive) integral. The product integral is also known as "continuous product" or "multiplical".
- Complex multiplication, a theory of elliptic curves.

3.10 See also

- Pi (letter)
- Iterated binary operation

3.11 Notes

[1] Here, "formal" means that this notation has the form of a determinant, but does not strictly adhere to the definition; it is a mnemonic used to remember the expansion of the cross product.

3.12 References

[1] Clarke, Francis (2013) *Functional analysis, calculus of variations and optimal control*. Dordrecht: Springer. pp. 9–10. ISBN 1447148207.

[2] Boothby, William M. (1986). *An introduction to differentiable manifolds and Riemannian geometry* (2nd ed.). Orlando: Academic Press. p. 200. ISBN 0080874398.

[3] Moschovakis, Yiannis (2006). *Notes on set theory* (2nd ed.). New York: Springer. p. 13. ISBN 0387316094.

3.13 External links

- Product on Wolfram Mathworld
- Product at PlanetMath.org.

Chapter 4

Variable (mathematics)

For variables in computer science, see Variable (computer science). For other uses, see Variable (disambiguation).

In elementary mathematics, a **variable** is an alphabetic character representing a number, called the **value** of the variable, which is either arbitrary or not fully specified or unknown. Making algebraic computations with variables as if they were explicit numbers allows one to solve a range of problems in a single computation. A typical example is the quadratic formula, which allows one to solve every quadratic equation by simply substituting the numeric values of the coefficients of the given equation to the variables that represent them.

The concept of **variable** is also fundamental in calculus. Typically, a function $y = f(x)$ involves two variables, y and x, representing respectively the value and the argument of the function. The term "variable" comes from the fact that, when the argument (also called the "variable of the function") *varies*, then the value *varies* accordingly.[1]

In more advanced mathematics, a **variable** is a symbol that denotes a mathematical object, which could be a number, a vector, a matrix, or even a function. In this case, the original property of "variability" of a variable is not kept (except, sometimes, for informal explanations).

Similarly, in computer science, a **variable** is a name (commonly an alphabetic character or a word) representing some value represented in computer memory. In mathematical logic, a **variable** is either a symbol representing an unspecified term of the theory, or a basic object of the theory, which is manipulated without referring to its possible intuitive interpretation.

4.1 Etymology

"Variable" comes from a Latin word, *variābilis*, with "*vari(us)*"' meaning "various" and "*-ābilis*"' meaning "-able", meaning "capable of changing".[2]

4.2 Genesis and evolution of the concept

François Viète introduced at the end of 16th century the idea of representing known and unknown numbers by letters, nowadays called variables, and of computing with them as if they were numbers, in order to obtain, at the end, the result by a simple replacement. François Viète's convention was to use consonants for known values and vowels for unknowns.[3]

In 1637, René Descartes "invented the convention of representing unknowns in equations by x, y, and z, and knowns by a, b, and c".[4] Contrarily to Viète's convention, Descartes' one is still commonly in use.

Starting in the 1660s, Isaac Newton and Gottfried Wilhelm Leibniz independently developed the infinitesimal calculus, which essentially consists of studying how an infinitesimal variation of a *variable quantity* induces a corresponding variation of another quantity which is a *function* of the first variable (quantity). Almost a century later Leonhard Euler fixed the terminology of infinitesimal calculus and introduced the notation $y = f(x)$ for a function f, its **variable** x and its value y. Until the end of the 19th century, the word *variable* referred almost exclusively to the arguments and the values of functions.

In the second half of the 19th century, it appeared that the foundation of infinitesimal calculus was not formalized enough to deal with apparent paradoxes such as a continuous function which is nowhere differentiable. To solve this problem, Karl Weierstrass introduced a new formalism consisting of replacing the intuitive notion of limit by a formal definition. The older notion of limit was "when the *variable* x varies and tends toward a, then $f(x)$ tends toward L", without any accurate definition of "tends". Weierstrass replaced this sentence by the formula

$$(\forall \epsilon > 0)(\exists \eta > 0)(\forall x) \; |x - a| < \eta \Rightarrow |L - f(x)| < \epsilon.$$

in which none of the five variables is considered as varying.

This static formulation led to the modern notion of variable which is simply a symbol representing a mathematical object which either is unknown or may be replaced by any element of a given set; for example, the set of real numbers.

4.3   Specific kinds of variables

It is common that many variables appear in the same mathematical formula, which play different roles. Some names or qualifiers have been introduced to distinguish them. For example, in the general cubic equation

$$ax^3 + bx^2 + cx + d = 0,$$

there are five variables. Four of them, a, b, c, d represent given numbers, and the last one, x, represents the *unknown* number, which is a solution of the equation. To distinguish them, the variable x is called *an unknown*, and the other variables are called *parameters* or *coefficients*, or sometimes *constants*, although this last terminology is incorrect for an equation and should be reserved for the function defined by the left-hand side of this equation.

In the context of functions, the term *variable* refers commonly to the arguments of the functions. This is typically the case in sentences like "function of a real variable", "x is the variable of the function $f: x \mapsto f(x)$", "f is a function of the variable x" (meaning that the argument of the function is referred to by the variable x).

In the same context, the variables that are independent of x define constant functions and are therefore called *constant*. For example, a *constant of integration* is an arbitrary constant function that is added to a particular antiderivative to obtain the other antiderivatives. Because the strong relationship between polynomials and polynomial function, the term "constant" is often used to denote the coefficients of a polynomial, which are constant functions of the indeterminates.

This use of "constant" as an abbreviation of "constant function" must be distinguished from the normal meaning of the word in mathematics. A **constant**, or **mathematical constant** is a well and unambiguously defined number or other mathematical object, as, for example, the numbers 0, 1, π and the identity element of a group.

Here are other specific names for variables.

- A **unknown** is a variable in which an equation has to be solved for.

- An **indeterminate** is a symbol, commonly called variable, that appears in a polynomial or a formal power series. Formally speaking, an indeterminate is not a variable, but a constant in the polynomial ring of the ring of formal power series. However, because of the strong relationship between polynomials or power series and the functions that they define, many authors consider indeterminates as a special kind of variables.

- A **parameter** is a quantity (usually a number) which is a part of the input of a problem, and remains constant during the whole solution of this problem. For example, in mechanics the mass and the size of a solid body are *parameters* for the study of its movement. It should be noted that in computer science, *parameter* has a different meaning and denotes an argument of a function.

- **Free variables and bound variables**

- A **random variable** is a kind of variable that is used in probability theory and its applications.

It should be emphasized that all these denominations of variables are of semantic nature and that the way of computing with them (syntax) is the same for all.

4.3.1   Dependent and independent variables

Main article: Dependent and independent variables

In calculus and its application to physics and other sciences, it is rather common to consider a variable, say y, whose possible values depend of the value of another variable, say x. In mathematical terms, the *dependent* variable y represents the value of a function of x. To simplify formulas, it is often useful to use the same symbol for the dependent variable y and the function mapping x onto y. For example, the state of a physical system depends on measurable quantities such as the pressure, the temperature, the spatial position, ..., and all these quantities varies when the system evolves, that is, they are function of the time. In the formulas describing the system, these quantities are represented by variables which are dependent on the time, and thus considered implicitly as functions of the time.

Therefore, in a formula, a **dependent variable** is a variable that is implicitly a function of another (or several other) variables. An **independent variable** is a variable that is not dependent.[5]

The property of a variable to be dependent or independent depends often of the point of view and is not intrinsic. For example, in the notation $f(x, y, z)$, the three variables may be all independent and the notation represents a function of three variables. On the other hand, if y and z depend on x (are *dependent variables*) then the notation represent a function of the single *independent variable x*.[6]

4.3.2   Examples

If one defines a function f from the real numbers to the real numbers by

$$f(x) = x^2 + \sin(x + 4)$$

then x is a variable standing for the argument of the function being defined, which can be any real number. In the identity

$$\sum_{i=1}^{n} i = \frac{n^2 + n}{2}$$

the variable i is a summation variable which designates in turn each of the integers 1, 2, ..., n (it is also called **index** because its variation is over a discrete set of values) while n is a parameter (it does not vary within the formula).

In the theory of polynomials, a polynomial of degree 2 is generally denoted as $ax^2 + bx + c$, where a, b and c are called coefficients (they are assumed to be fixed, i.e., parameters of the problem considered) while x is called a variable. When studying this polynomial for its polynomial function this x stands for the function argument. When studying the polynomial as an object in itself, x is taken to be an indeterminate, and would often be written with a capital letter instead to indicate this status.

4.4   Notation

In mathematics, the variables are generally denoted by a single letter. However, this letter is frequently followed by a subscript, as in x_2, and this subscript may be a number, another variable (x_i), a word or the abbreviation of a word (x_{in}

and x_{out}), and even a mathematical expression. Under the influence of computer science, one may encounter in pure mathematics some variable names consisting in several letters and digits.

Following the 17th century French philosopher and mathematician, René Descartes, letters at the beginning of the alphabet, e.g. a, b, c are commonly used for known values and parameters, and letters at the end of the alphabet, e.g. x, y, z, and t are commonly used for unknowns and variables of functions.[7] In printed mathematics, the norm is to set variables and constants in an italic typeface.[8]

For example, a general quadratic function is conventionally written as:

$$ax^2 + bx + c,$$

where a, b and c are parameters (also called constants, because they are constant functions), while x is the variable of the function. A more explicit way to denote this function is

$$x \mapsto ax^2 + bx + c,$$

which makes the function-argument status of x clear, and thereby implicitly the constant status of a, b and c. Since c occurs in a term that is a constant function of x, it is called the constant term.[9]:18

Specific branches and applications of mathematics usually have specific naming conventions for variables. Variables with similar roles or meanings are often assigned consecutive letters. For example, the three axes in 3D coordinate space are conventionally called x, y, and z. In physics, the names of variables are largely determined by the physical quantity they describe, but various naming conventions exist. A convention often followed in probability and statistics is to use X, Y, Z for the names of random variables, keeping x, y, z for variables representing corresponding actual values.

There are many other notational usages. Usually, variables that play a similar role are represented by consecutive letters or by the same letter with different subscript. Below are some of the most common usages.

- a, b, c, and d (sometimes extended to e and f) often represent parameters or coefficients.

- a_0, a_1, a_2, ... play a similar role, when otherwise too many different letters would be needed.

- ai or ui is often used to denote the i-th term of a sequence or the i-th coefficient of a series.

- f and g (sometimes h) commonly denote functions.

- i, j, and k (sometimes l or h) are often used to denote varying integers or indices in an indexed family.

- l and w are often used to represent the length and width of a figure.

- l is also used to denote a line. In number theory, l often denotes a prime number not equal to p.

- n usually denotes a fixed integer, such as a count of objects or the degree of an equation.

  - When two integers are needed, for example for the dimensions of a matrix, one uses commonly m and n.

- p often denotes a prime numbers or a probability.

- q often denotes a prime power or a quotient

- r often denotes a remainder.

- t often denotes time.

- x, y and z usually denote the three Cartesian coordinates of a point in Euclidean geometry. By extension, they are used to name the corresponding axes.

- z typically denotes a complex number, or, in statistics, a normal random variable.

- $\alpha, \beta, \gamma, \theta$ and φ commonly denote angle measures.

- ε usually represents an arbitrarily small positive number.

  - ε and δ commonly denote two small positives.

- λ is used for eigenvalues.

- σ often denotes a sum, or, in statistics, the standard deviation.

4.5   See also

- Free variables and bound variables (Bound variables are also known as dummy variables)

- Variable (programming)

- Mathematical expression

- Physical constant

- Coefficient

- Constant of integration

- Constant term of a polynomial

- Indeterminate (variable)

- Lambda calculus

4.6   Bibliography

- J. Edwards (1892). *Differential Calculus*. London: MacMillan and Co. pp. 1 ff.

- Karl Menger, "On Variables in Mathematics and in Natural Science", *The British Journal for the Philosophy of Science* **5**:18:134-142 (August 1954) JSTOR 685170

- Jaroslav Peregrin, "Variables in Natural Language: Where do they come from?", in M. Boettner, W. Thümmel, eds., *Variable-Free Semantics*, 2000, p. 46-65.

- W. V. Quine, "Variables Explained Away", *Proceedings of the American Philosophical Society* **104**:343-347 (1960).

4.7   References

[1] *Syracuse University*. "Appendix One Review of Constants and Variables". cstl.syr.edu.

[2] ""Variable" Origin". dictionary.com. Retrieved 18 May 2015.

[3] Fraleigh, John B. (1989). *A First Course in Abstract Algebra* (4 ed.). United States: Addison-Wesley. p. 276. ISBN 0-201-52821-5.

[4] Tom Sorell, *Descartes: A Very Short Introduction*, (2000). New York: Oxford University Press. p. 19.

[5] Edwards Art. 5

[6] Edwards Art. 6

[7] Edwards Art. 4

[8] William L. Hosch (editor), *The Britannica Guide to Algebra and Trigonometry*, Britannica Educational Publishing, The Rosen Publishing Group, 2010, ISBN 1615302190, 9781615302192, page 71

[9] Foerster, Paul A. (2006). *Algebra and Trigonometry: Functions and Applications, Teacher's Edition* (Classics ed.). Upper Saddle River, NJ: Prentice Hall. ISBN 0-13-165711-9.

Chapter 5

Coefficient

For other uses, see Coefficient (disambiguation).

In mathematics, a **coefficient** is a multiplicative factor in some term of a polynomial, a series or any expression; it is usually a number, but in any case does not involve any variables of the expression. For instance in

$$7x^2 - 3xy + 1.5 + y$$

the first two terms respectively have the coefficients 7 and −3. The third term 1.5 is a constant. The final term does not have any explicitly written coefficient, but is considered to have coefficient 1, since multiplying by that factor would not change the term. Often coefficients are numbers as in this example, although they could be parameters of the problem, as a, b, and c, where "c" is a constant, in

$$ax^2 + bx + c$$

when it is understood that these are not considered variables. Thus a polynomial in one variable x can be written as

$$a_k x^k + \cdots + a_1 x^1 + a_0$$

for some integer k, where $a_k, \ldots, a_1, a_0$ are coefficients; to allow this kind of expression in all cases one must allow introducing terms with 0 as coefficient. For the largest i with $a_i \neq 0$ (if any), a_i is called the **leading coefficient** of the polynomial. So for example the leading coefficient of the polynomial

$$4x^5 + x^3 + 2x^2$$

is 4.

Specific coefficients arise in mathematical identities, such as the binomial theorem which involves binomial coefficients; these particular coefficients are tabulated in Pascal's triangle.

5.1 Linear algebra

In linear algebra, the **leading coefficient** of a row in a matrix is the first nonzero entry in that row. So, for example, given

$$M = \begin{pmatrix} 1 & 2 & 0 & 6 \\ 0 & 2 & 9 & 4 \\ 0 & 0 & 0 & 4 \\ 0 & 0 & 0 & 0 \end{pmatrix}$$

The leading coefficient of the first row is 1; 2 is the leading coefficient of the second row; 4 is the leading coefficient of the third row, and the last row does not have a leading coefficient.

Though coefficients are frequently viewed as constants in elementary algebra, they can be variables more generally. For example, the coordinates $(x_1, x_2, \ldots, x_n)$ of a vector v in a vector space with basis $\{e_1, e_2, \ldots, e_n\}$, are the coefficients of the basis vectors in the expression

$$v = x_1 e_1 + x_2 e_2 + \cdots + x_n e_n.$$

5.2 Examples of physical coefficients

1. *Coefficient of Thermal Expansion* (thermodynamics) (dimensionless) - Relates the change in temperature to the change in a material's dimensions.

2. *Partition Coefficient* (*KD*) (chemistry) - The ratio of concentrations of a compound in two phases of a mixture of two immiscible solvents at equilibrium. H2O is a coefficient

3. *Hall coefficient* (electrical physics) - Relates a magnetic field applied to an element to the voltage created, the amount of current and the element thickness. It is a characteristic of the material from which the conductor is made.

4. *Lift coefficient* (*CL* or *CZ*) (Aerodynamics) (dimensionless) - Relates the lift generated by an airfoil with the dynamic pressure of the fluid flow around the airfoil, and the plan-form area of the airfoil.

5. *Ballistic coefficient* (BC) (Aerodynamics) (units of kg/m^2) - A measure of a body's ability to overcome air resistance in flight. BC is a function of mass, diameter, and drag coefficient.

6. *Transmission Coefficient* (quantum mechanics) (dimensionless) - Represents the probability flux of a transmitted wave relative to that of an incident wave. It is often used to describe the probability of a particle tunnelling through a barrier.

7. *Damping Factor* a.k.a. *viscous damping coefficient* (Physical Engineering) (units of newton-seconds per meter) - relates a damping force with the velocity of the object whose motion is being damped.

A coefficient is a number placed in front of a term in a chemical equation to indicate how many molecules (or atoms) take part in the reaction. For example, in the formula

$$2H_2 + O_2 \rightarrow 2H_2O$$

the number 2's in front of H_2 and H_2O are stoichiometric coefficients.

5.3 See also

- Degree of a polynomial

- Monic polynomial

5.4   References

- Sabah Al-hadad and C.H. Scott (1979) *College Algebra with Applications*, page 42, Winthrop Publishers, Cambridge Massachusetts ISBN 0-87626-140-3 .

- Gordon Fuller, Walter L Wilson, Henry C Miller, (1982) *College Algebra*, 5th edition, page 24, Brooks/Cole Publishing, Monterey California ISBN 0-534-01138-1 .

- Steven Schwartzman (1994) *The Words of Mathematics: an etymological dictionary of mathematical terms used in English*, page 48, Mathematics Association of America, ISBN 0-88385-511-9.

Chapter 6

Ratio

For non-dimensionless ratios, see Rates.

For other uses, see Ratio (disambiguation).

"is to" redirects here. For the grammatical construction, see am to.

In mathematics, a **ratio** is a relationship between two numbers indicating how many times the first number contains the

The ratio of width to height of standard-definition television.

second.[1] For example, if a bowl of fruit contains eight oranges and six lemons, then the ratio of oranges to lemons is eight to six (that is, 8:6, which is equivalent to the ratio 4:3). Thus, a ratio can be a fraction as opposed to a whole number.

Also, in this example the ratio of lemons to oranges is 6:8 (or 3:4), and the ratio of oranges to the total amount of fruit is 8:14 (or 4:7).

The numbers compared in a ratio can be any quantities of a comparable kind, such as objects, persons, lengths, or spoonfuls. A ratio is written "*a* to *b*" or *a:b*, or sometimes expressed arithmetically as a quotient of the two.[2] When the two quantities have the same units, as is often the case, their ratio is a dimensionless number. A rate is a quotient of variables having different units. But in many applications, the word *ratio* is often used instead for this more general notion as well.[3]

6.1 Notation and terminology

The ratio of numbers *A* and *B* can be expressed as:[4]

- the ratio of **A** to **B**

- **A** is to **B** (*followed by* "*as* C *is to* D")

- **A:B**

- A fraction that is the quotient: **A** divided by **B**: $\frac{A}{B}$, which can be expressed as either a simple or a decimal fraction.[5]

The numbers *A* and *B* are sometimes called *terms* with *A* being the *antecedent* and *B* being the *consequent*.[6]

The proportion expressing the equality of the ratios *A:B* and *C:D* is written *A:B = C:D* or *A:B::C:D*. This latter form, when spoken or written in the English language, is often expressed as

> *A* is to *B* as *C* is to *D*.

A, *B*, *C* and *D* are called the terms of the proportion. *A* and *D* are called the *extremes*, and *B* and *C* are called the *means*. The equality of three or more proportions is called a continued proportion.[7]

Ratios are sometimes used with three or more terms. The ratio of the dimensions of a "two by four" that is ten inches long is 2:4:10. A good concrete mix is sometimes quoted as 1:2:4 for the ratio of cement to sand to gravel.[8]

For a mixture of 4/1 cement to water, it could be said that the ratio of cement to water is 4:1, that there is 4 times as much cement as water, or that there is a quarter (1/4) as much water as cement.

Older televisions have a 4:3 *aspect ratio*, which means that the width is 4/3 of the height (this can also be expressed as 1.33:1 or just 1.33 rounded to two decimal places); modern widescreen TVs have a 16:9 (or 1.78 rounded to two decimal places) aspect ratio. One of the popular widescreen movie formats is 2.35:1 or simply 2.35. Representing ratios as decimal fractions simplifies their comparison.

6.2 History and etymology

It is impossible to trace the origin of the *concept* of ratio, because the ideas from which it developed would have been familiar to preliterate cultures. For example, the idea of one village being twice as large as another is so basic that it would have been understood in prehistoric society.[9] However, it is possible to trace the origin of the word "ratio" to the Ancient Greek λόγος (*logos*). Early translators rendered this into Latin as *ratio* ("reason"; as in the word "rational"). (A rational number may be expressed as the quotient of two integers.) A more modern interpretation of Euclid's meaning is more akin to computation or reckoning.[10] Medieval writers used the word *proportio* ("proportion") to indicate ratio and *proportionalitas* ("proportionality") for the equality of ratios.[11]

Euclid collected the results appearing in the Elements from earlier sources. The Pythagoreans developed a theory of ratio and proportion as applied to numbers.[12] The Pythagoreans' conception of number included only what would today be called rational numbers, casting doubt on the validity of the theory in geometry where, as the Pythagoreans also discovered, incommensurable ratios (corresponding to irrational numbers) exist. The discovery of a theory of ratios that does not assume commensurability is probably due to Eudoxus of Cnidus. The exposition of the theory of proportions that appears in Book VII of The Elements reflects the earlier theory of ratios of commensurables.[13]

The existence of multiple theories seems unnecessarily complex to modern sensibility since ratios are, to a large extent, identified with quotients. This is a comparatively recent development however, as can be seen from the fact that modern geometry textbooks still use distinct terminology and notation for ratios and quotients. The reasons for this are twofold. First, there was the previously mentioned reluctance to accept irrational numbers as true numbers. Second, the lack of a widely used symbolism to replace the already established terminology of ratios delayed the full acceptance of fractions as alternative until the 16th century.[14]

6.2.1 Euclid's definitions

Book V of Euclid's Elements has 18 definitions, all of which relate to ratios.[15] In addition, Euclid uses ideas that were in such common usage that he did not include definitions for them. The first two definitions say that a *part* of a quantity is another quantity that "measures" it and conversely, a *multiple* of a quantity is another quantity that it measures. In modern terminology, this means that a multiple of a quantity is that quantity multiplied by an integer greater than one—and a part of a quantity (meaning aliquot part) is a part that, when multiplied by an integer greater than one, gives the quantity.

Euclid does not define the term "measure" as used here. However, one may infer that if a quantity is taken as a unit of measurement, and a second quantity is given as an integral number of these units, then the first quantity *measures* the second. Note that these definitions are repeated, nearly word for word, as definitions 3 and 5 in book VII.

Definition 3 describes what a ratio is in a general way. It is not rigorous in a mathematical sense and some have ascribed it to Euclid's editors rather than Euclid himself.[16] Euclid defines a ratio as between two quantities *of the same type*, so by this definition the ratios of two lengths or of two areas are defined, but not the ratio of a length and an area. Definition 4 makes this more rigorous. It states that a ratio of two quantities exists when there is a multiple of each that exceeds the other. In modern notation, a ratio exists between quantities p and q if there exist integers m and n so that $mp>q$ and $nq>p$. This condition is known as the Archimedes property.

Definition 5 is the most complex and difficult. It defines what it means for two ratios to be equal. Today, this can be done by simply stating that ratios are equal when the quotients of the terms are equal, but Euclid did not accept the existence of the quotients of incommensurate, so such a definition would have been meaningless to him. Thus, a more subtle definition is needed where quantities involved are not measured directly to one another. Though it may not be possible to assign a rational value to a ratio, it is possible to compare a ratio with a rational number. Specifically, given two quantities, p and q, and a rational number m/n we can say that the ratio of p to q is less than, equal to, or greater than m/n when np is less than, equal to, or greater than mq respectively. Euclid's definition of equality can be stated as that two ratios are equal when they behave identically with respect to being less than, equal to, or greater than any rational number. In modern notation this says that given quantities p, q, r and s, then $p{:}q{:}{:}r{:}s$ if for any positive integers m and n, $np<mq$, $np=mq$, $np>mq$ according as $nr<ms$, $nr=ms$, $nr>ms$ respectively. There is a remarkable similarity between this definition and the theory of Dedekind cuts used in the modern definition of irrational numbers.[17]

Definition 6 says that quantities that have the same ratio are *proportional* or *in proportion*. Euclid uses the Greek ἀναλόγον (analogon), this has the same root as λόγος and is related to the English word "analog".

Definition 7 defines what it means for one ratio to be less than or greater than another and is based on the ideas present in definition 5. In modern notation it says that given quantities p, q, r and s, then $p{:}q>r{:}s$ if there are positive integers m and n so that $np>mq$ and $nr\leq ms$.

As with definition 3, definition 8 is regarded by some as being a later insertion by Euclid's editors. It defines three terms p, q and r to be in proportion when $p{:}q{:}{:}q{:}r$. This is extended to 4 terms p, q, r and s as $p{:}q{:}{:}q{:}r{:}{:}r{:}s$, and so on. Sequences that have the property that the ratios of consecutive terms are equal are called geometric progressions. Definitions 9 and 10 apply this, saying that if p, q and r are in proportion then $p{:}r$ is the *duplicate ratio* of $p{:}q$ and if p, q, r and s are in proportion then $p{:}s$ is the *triplicate ratio* of $p{:}q$. If p, q and r are in proportion then q is called a *mean proportional* to (or the geometric mean of) p and r. Similarly, if p, q, r and s are in proportion then q and r are called two mean proportionals to p and s.

6.3   Number of terms and use of fractions

In general, a comparison of the quantities of a two-entity ratio can be expressed as a fraction derived from the ratio. For example, in a ratio of 2:3, the amount, size, volume, or quantity of the first entity is $\frac{2}{3}$ that of the second entity.

If there are 2 oranges and 3 apples, the ratio of oranges to apples is 2:3, and the ratio of oranges to the total number of pieces of fruit is 2:5. These ratios can also be expressed in fraction form: there are 2/3 as many oranges as apples, and 2/5 of the pieces of fruit are oranges. If orange juice concentrate is to be diluted with water in the ratio 1:4, then one part of concentrate is mixed with four parts of water, giving five parts total; the amount of orange juice concentrate is 1/4 the amount of water, while the amount of orange juice concentrate is 1/5 of the total liquid. In both ratios and fractions, it is important to be clear what is being compared to what, and beginners often make mistakes for this reason.

Fractions can also be inferred from ratios with more than two entities; however, a ratio with more than two entities cannot be completely converted into a single fraction, because a fraction can only compare two quantities. A separate fraction can be used to compare the quantities of any two of the entities covered by the ratio: for example, from a ratio of 2:3:7 we can infer that the quantity of the second entity is $\frac{3}{7}$ that of the third entity.

6.4   Proportions and percentage ratios

If we multiply all quantities involved in a ratio by the same number, the ratio remains valid. For example, a ratio of 3:2 is the same as 12:8. It is usual either to reduce terms to the lowest common denominator, or to express them in parts per hundred (percent).

If a mixture contains substances A, B, C and D in the ratio 5:9:4:2 then there are 5 parts of A for every 9 parts of B, 4 parts of C and 2 parts of D. As 5+9+4+2=20, the total mixture contains 5/20 of A (5 parts out of 20), 9/20 of B, 4/20 of C, and 2/20 of D. If we divide all numbers by the total and multiply by 100%, we have converted to percentages: 25% A, 45% B, 20% C, and 10% D (equivalent to writing the ratio as 25:45:20:10).

If the two or more ratio quantities encompass all of the quantities in a particular situation, it is said that "the whole" contains the sum of the parts: for example, a fruit basket containing two apples and three oranges and no other fruit is made up of two parts apples and three parts oranges. In this case, $\frac{2}{5}$, or 40% of the whole is apples and $\frac{3}{5}$, or 60% of the whole is oranges. This comparison of a specific quantity to "the whole" is called a proportion.

6.5   Reduction

Ratios can be reduced (as fractions are) by dividing each quantity by the common factors of all the quantities. As for fractions, the simplest form is considered that in which the numbers in the ratio are the smallest possible integers.

Thus, the ratio 40:60 is equivalent in meaning to the ratio 2:3, the latter being obtained from the former by dividing both quantities by 20. Mathematically, we write 40:60 = 2:3, or equivalently 40:60::2:3. The verbal equivalent is "40 is to 60 as 2 is to 3."

A ratio that has integers for both quantities and that cannot be reduced any further (using integers) is said to be in simplest form or lowest terms.

Sometimes it is useful to write a ratio in the form 1:x or x:1, where x is not necessarily an integer, to enable comparisons of different ratios. For example, the ratio 4:5 can be written as 1:1.25 (dividing both sides by 4) Alternatively, it can be written as 0.8:1 (dividing both sides by 5).

Where the context makes the meaning clear, a ratio in this form is sometimes written without the 1 and the colon, though, mathematically, this makes it a factor or multiplier.

6.6 Irrational ratios

Some ratios are between incommensurable quantities—quantities whose ratio is an irrational number. The earliest discovered example, found by the Pythagoreans, is the ratio of the diagonal to the side of a square, which is the square root of 2.

The ratio of a circle's circumference to its diameter is called pi, and is not only irrational but also transcendental.

Another well-known example is the golden ratio, which is defined as both sides of the equality $a{:}b = (a+b){:}a$. Writing this in fractional terms as $(a/b) = 1 + \frac{1}{(a/b)}$ and finding the positive solution gives the golden ratio $\frac{a}{b} = \frac{1+\sqrt{5}}{2}$, which is irrational. Thus at least one of a and b has to be irrational for them to be in the golden ratio. An example of an occurrence of the golden ratio is as the limiting value of the ratio of two successive Fibonacci numbers: even though the n-th such ratio is the ratio of two integers and hence is rational, the limit of the sequence of these ratios as n goes to infinity is the irrational golden ratio.

Similarly, the silver ratio is defined as both sides of the equality $a{:}b = (2a+b){:}a$. Again writing it in fractional terms and obtaining the positive solution, we obtain $\frac{a}{b} = 1 + \sqrt{2}$, which is irrational, so of two quantities a and b in the silver ratio at least one of them must be irrational.

6.7 Odds

Main article: Odds

Odds (as in gambling) are expressed as a ratio. For example, odds of "7 to 3 against" (7:3) mean that there are seven chances that the event will not happen to every three chances that it will happen. The probability of success is 30%. In every ten trials, there are expected to be three wins and seven losses.

6.8 Units

Ratios may be unitless, as in the case they relate quantities in units of the same dimension, even in their units of measurement are initially different. For example, the ratio 1 minute : 40 seconds can be reduced by changing the first value to 60 seconds. Once the units are the same, they can be omitted, and the ratio can be reduced to 3:2.

On the other hand, there are non-dimensionless ratios, also known as rates.[18][19] In chemistry, mass concentration ratios are usually expressed as weight/volume fractions. For example, a concentration of 3% w/v usually means 3g of substance in every 100mL of solution. This cannot be converted to a dimensionless ratio, as in weight/weight or volume/volume fractions.

6.9 Triangular coordinates

The locations of points relative to a triangle with vertices A, B, and C and sides AB, BC, and CA are often expressed in extended ratio form as *triangular coordinates*.

In barycentric coordinates, a point with coordinates $\alpha : \beta : \gamma$ is the point upon which a weightless sheet of metal in the shape and size of the triangle would exactly balance if weights were put on the vertices, with the ratio of the weights at A and B being $\alpha : \beta$, the ratio of the weights at B and C being $\beta : \gamma$, and therefore the ratio of weights at A and C being $\alpha : \gamma$.

In trilinear coordinates, a point with coordinates $x{:}y{:}z$ has perpendicular distances to side BC (across from vertex A) and side CA (across from vertex B) in the ratio $x{:}y$, distances to side CA and side AB (across from C) in the ratio $y{:}z$, and therefore distances to sides BC and AB in the ratio $x{:}z$.

Since all information is expressed in terms of ratios (the individual numbers denoted by $\alpha, \beta, \gamma, x, y,$ and z have no meaning by themselves), a triangle analysis using barycentric or trilinear coordinates applies regardless of the size of the triangle.

6.10   See also

- Dilution ratio
- Dimensionless quantity
- Financial ratio
- Fold change
- Interval (music)
- Odds ratio
- Parts-per notation
- Price–performance ratio
- Proportionality (mathematics)
- Ratio distribution
- Ratio estimator
- Rate (mathematics)
- Rate ratio
- Relative risk
- Rule of three (mathematics)
- Sex ratio
- Slope

6.11   References

[1] Penny Cyclopedia, p. 307

[2] New International Encyclopedia

[3] *"The quotient of two numbers (or quantities); the relative sizes of two numbers (or quantities)"*, "The Mathematics Dictionary"

[4] New International Encyclopedia

[5] Decimal fractions are frequently used in technological areas where ratio comparisons are important, such as aspect ratios (imaging), compression ratios (engines or data storage), etc.

[6] from the Encyclopedia Britannica

[7] New International Encyclopedia

[8] Belle Group concrete mixing hints

[9] Smith, p. 477

[10] Penny Cyclopedia, p. 307

[11] Smith, p. 478

[12] Heath, p. 112

[13] Heath, p. 113

[14] Smith, p. 480

[15] Heath, reference for section

[16] "Geometry, Euclidean" *Encyclopædia Britannica Eleventh Edition* p682.

[17] Heath p. 125

[18] "'Velocity' can be defined as the ratio... 'Population density' is the ratio... 'Gasoline consumption' is measure as the ratio...", "Ratio and Proportion: Research and Teaching in Mathematics Teachers"

[19] "Ratio as a Rate. *The first type [of ratio] defined by Freudenthal, above, is known as rate, and illustrates a comparison between two variables with difference units. (...) A ratio of this sort produces a unique, new concept with its own entity, and this new concept is usually not considered a ratio, per se, but a rate or density.*", "Ratio and Proportion: Research and Teaching in Mathematics Teachers"

6.12   Further reading

- "Ratio" *The Penny Cyclopædia* vol. 19, The Society for the Diffusion of Useful Knowledge (1841) Charles Knight and Co., London pp. 307ff

- "Proportion" *New International Encyclopedia, Vol. 19* 2nd ed. (1916) Dodd Mead & Co. pp270-271

- "Ratio and Proportion" *Fundamentals of practical mathematics*, George Wentworth, David Eugene Smith, Herbert Druery Harper (1922) Ginn and Co. pp. 55ff

- *The thirteen books of Euclid's Elements, vol 2.* trans. Sir Thomas Little Heath (1908). Cambridge Univ. Press. pp. 112ff.

- D.E. Smith, *History of Mathematics, vol 2* Dover (1958) pp. 477ff

6.13   External links

Chapter 7

Correlation and dependence

This article is about correlation and dependence in statistical data. For other uses, see correlation (disambiguation).

In statistics, **dependence** is any statistical relationship between two random variables or two sets of data. **Correlation** refers to any of a broad class of statistical relationships involving dependence. Familiar examples of dependent phenomena include the correlation between the physical statures of parents and their offspring, and the correlation between the demand for a product and its price.

Correlations are useful because they can indicate a predictive relationship that can be exploited in practice. For example, an electrical utility may produce less power on a mild day based on the correlation between electricity demand and weather. In this example there is a causal relationship, because extreme weather causes people to use more electricity for heating or cooling; however, statistical dependence is not sufficient to demonstrate the presence of such a causal relationship (i.e., correlation does not imply causation).

Formally, *dependence* refers to any situation in which random variables do not satisfy a mathematical condition of probabilistic independence. In loose usage, *correlation* can refer to any departure of two or more random variables from independence, but technically it refers to any of several more specialized types of relationship between mean values. There are several **correlation coefficients**, often denoted ϱ or r, measuring the degree of correlation. The most common of these is the Pearson correlation coefficient, which is sensitive only to a linear relationship between two variables (which may exist even if one is a nonlinear function of the other). Other correlation coefficients have been developed to be more robust than the Pearson correlation – that is, more sensitive to nonlinear relationships.[1][2][3] Mutual information can also be applied to measure dependence between two variables.

7.1 Pearson's product-moment coefficient

Main article: Pearson product-moment correlation coefficient

The most familiar measure of dependence between two quantities is the Pearson product-moment correlation coefficient, or "Pearson's correlation coefficient", commonly called simply "the correlation coefficient". It is obtained by dividing the covariance of the two variables by the product of their standard deviations. Karl Pearson developed the coefficient from a similar but slightly different idea by Francis Galton.[4]

The population correlation coefficient $\rho_{X,Y}$ between two random variables X and Y with expected values μ_X and μ_Y and standard deviations σ_X and σ_Y is defined as:

$$\rho_{X,Y} = \operatorname{corr}(X,Y) = \frac{\operatorname{cov}(X,Y)}{\sigma_X \sigma_Y} = \frac{E[(X - \mu_X)(Y - \mu_Y)]}{\sigma_X \sigma_Y}.$$

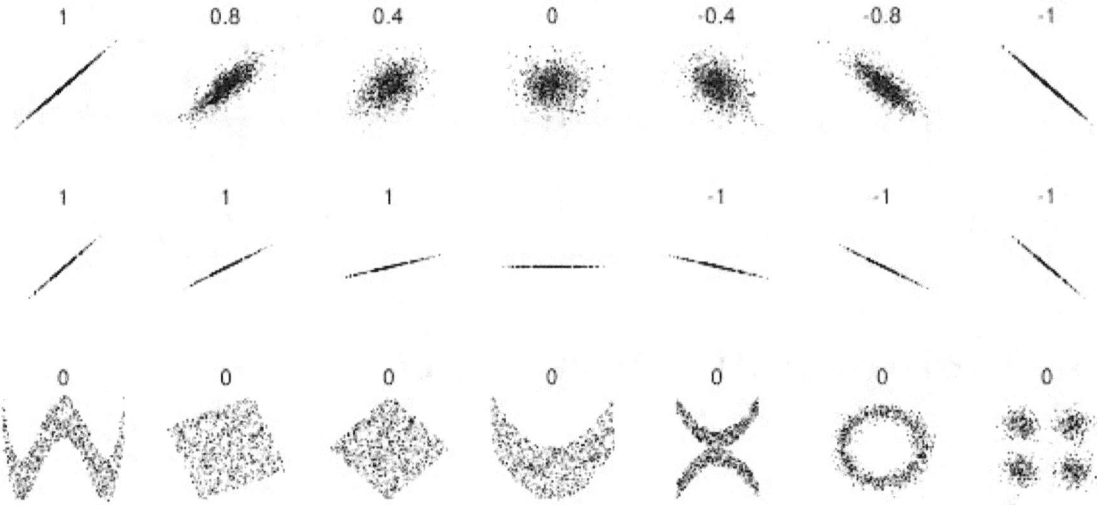

Several sets of (x, y) points, with the Pearson correlation coefficient of x and y for each set. Note that the correlation reflects the noisiness and direction of a linear relationship (top row), but not the slope of that relationship (middle), nor many aspects of nonlinear relationships (bottom). N.B.: the figure in the center has a slope of 0 but in that case the correlation coefficient is undefined because the variance of Y is zero.

where E is the expected value operator, *cov* means covariance, and *corr* is a widely used alternative notation for the correlation coefficient.

The Pearson correlation is defined only if both of the standard deviations are finite and nonzero. It is a corollary of the Cauchy–Schwarz inequality that the correlation cannot exceed 1 in absolute value. The correlation coefficient is symmetric: $\mathrm{corr}(X, Y) = \mathrm{corr}(Y, X)$.

The Pearson correlation is +1 in the case of a perfect direct (increasing) linear relationship (correlation), −1 in the case of a perfect decreasing (inverse) linear relationship (**anticorrelation**),[5] and some value between −1 and 1 in all other cases, indicating the degree of linear dependence between the variables. As it approaches zero there is less of a relationship (closer to uncorrelated). The closer the coefficient is to either −1 or 1, the stronger the correlation between the variables.

If the variables are independent, Pearson's correlation coefficient is 0, but the converse is not true because the correlation coefficient detects only linear dependencies between two variables. For example, suppose the random variable X is symmetrically distributed about zero, and $Y = X^2$. Then Y is completely determined by X, so that X and Y are perfectly dependent, but their correlation is zero; they are uncorrelated. However, in the special case when X and Y are jointly normal, uncorrelatedness is equivalent to independence.

If we have a series of n measurements of X and Y written as x_i and y_i where $i = 1, 2, \ldots, n$, then the *sample correlation coefficient* can be used to estimate the population Pearson correlation r between X and Y. The sample correlation coefficient is written

$$r_{xy} = \frac{\sum_{i=1}^{n}(x_i - \bar{x})(y_i - \bar{y})}{(n-1)s_x s_y} = \frac{\sum_{i=1}^{n}(x_i - \bar{x})(y_i - \bar{y})}{\sqrt{\sum_{i=1}^{n}(x_i - \bar{x})^2 \sum_{i=1}^{n}(y_i - \bar{y})^2}}.$$

where x and y are the sample means of X and Y, and s_x and s_y are the sample standard deviations of X and Y.

This can also be written as:

$$r_{xy} = \frac{\sum x_i y_i - n\bar{x}\bar{y}}{(n-1)s_x s_y} = \frac{n\sum x_i y_i - \sum x_i \sum y_i}{\sqrt{n\sum x_i^2 - (\sum x_i)^2}\sqrt{n\sum y_i^2 - (\sum y_i)^2}}.$$

If x and y are results of measurements that contain measurement error, the realistic limits on the correlation coefficient are not −1 to +1 but a smaller range.[6]

For the case of a linear model with a single independent variable, the coefficient of determination (R squared) is the square of r, Pearson's product-moment coefficient .

7.2   Rank correlation coefficients

Main articles: Spearman's rank correlation coefficient and Kendall tau rank correlation coefficient

Rank correlation coefficients, such as Spearman's rank correlation coefficient and Kendall's rank correlation coefficient (τ) measure the extent to which, as one variable increases, the other variable tends to increase, without requiring that increase to be represented by a linear relationship. If, as the one variable increases, the other *decreases*, the rank correlation coefficients will be negative. It is common to regard these rank correlation coefficients as alternatives to Pearson's coefficient, used either to reduce the amount of calculation or to make the coefficient less sensitive to non-normality in distributions. However, this view has little mathematical basis, as rank correlation coefficients measure a different type of relationship than the Pearson product-moment correlation coefficient, and are best seen as measures of a different type of association, rather than as alternative measure of the population correlation coefficient.[7][8]

To illustrate the nature of rank correlation, and its difference from linear correlation, consider the following four pairs of numbers (x, y):

   (0, 1), (10, 100), (101, 500), (102, 2000).

As we go from each pair to the next pair x increases, and so does y. This relationship is perfect, in the sense that an increase in x is *always* accompanied by an increase in y. This means that we have a perfect rank correlation, and both Spearman's and Kendall's correlation coefficients are 1, whereas in this example Pearson product-moment correlation coefficient is 0.7544, indicating that the points are far from lying on a straight line. In the same way if y always *decreases* when x *increases*, the rank correlation coefficients will be −1, while the Pearson product-moment correlation coefficient may or may not be close to −1, depending on how close the points are to a straight line. Although in the extreme cases of perfect rank correlation the two coefficients are both equal (being both +1 or both −1), this is not generally the case, and so values of the two coefficients cannot meaningfully be compared.[7] For example, for the three pairs (1, 1) (2, 3) (3, 2) Spearman's coefficient is 1/2, while Kendall's coefficient is 1/3.

7.3   Other measures of dependence among random variables

The information given by a correlation coefficient is not enough to define the dependence structure between random variables.[9] The correlation coefficient completely defines the dependence structure only in very particular cases, for example when the distribution is a multivariate normal distribution. (See diagram above.) In the case of elliptical distributions it characterizes the (hyper-)ellipses of equal density, however, it does not completely characterize the dependence structure (for example, a multivariate t-distribution's degrees of freedom determine the level of tail dependence).

Distance correlation and Brownian covariance / Brownian correlation[10][11] were introduced to address the deficiency of Pearson's correlation that it can be zero for dependent random variables; zero distance correlation and zero Brownian correlation imply independence.

The Randomized Dependence Coefficient[12] is a computationally efficient, copula-based measure of dependence between multivariate random variables. RDC is invariant with respect to non-linear scalings of random variables, is capable of discovering a wide range of functional association patterns and takes value zero at independence.

The correlation ratio is able to detect almost any functional dependency, and the entropy-based mutual information, total correlation and dual total correlation are capable of detecting even more general dependencies. These are sometimes referred to as multi-moment correlation measures, in comparison to those that consider only second moment (pairwise or quadratic) dependence.

The polychoric correlation is another correlation applied to ordinal data that aims to estimate the correlation between theorised latent variables.

One way to capture a more complete view of dependence structure is to consider a copula between them.

The coefficient of determination generalizes the correlation coefficient for relationships beyond simple linear regression.

7.4 Sensitivity to the data distribution

The degree of dependence between variables X and Y does not depend on the scale on which the variables are expressed. That is, if we are analyzing the relationship between X and Y, most correlation measures are unaffected by transforming X to $a + bX$ and Y to $c + dY$, where a, b, c, and d are constants (b and d being positive). This is true of some correlation statistics as well as their population analogues. Some correlation statistics, such as the rank correlation coefficient, are also invariant to monotone transformations of the marginal distributions of X and/or Y.

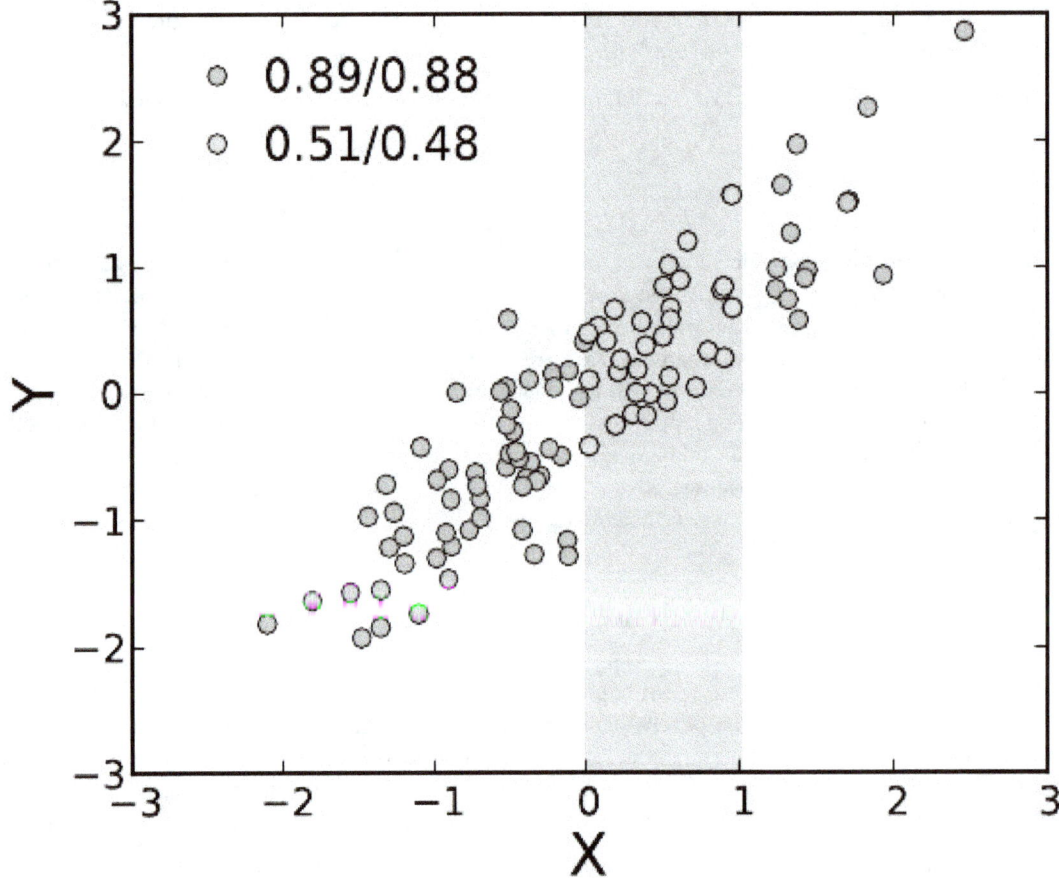

Pearson/Spearman correlation coefficients between X *and* Y *are shown when the two variables' ranges are unrestricted, and when the range of* X *is restricted to the interval (0.1).*

Most correlation measures are sensitive to the manner in which X and Y are sampled. Dependencies tend to be stronger if viewed over a wider range of values. Thus, if we consider the correlation coefficient between the heights of fathers and their sons over all adult males, and compare it to the same correlation coefficient calculated when the fathers are selected to be between 165 cm and 170 cm in height, the correlation will be weaker in the latter case. Several techniques have been developed that attempt to correct for range restriction in one or both variables, and are commonly used in meta-analysis: the most common are Thorndike's case II and case III equations.[13]

Various correlation measures in use may be undefined for certain joint distributions of X and Y. For example, the Pearson correlation coefficient is defined in terms of moments, and hence will be undefined if the moments are undefined. Measures of dependence based on quantiles are always defined. Sample-based statistics intended to estimate population measures of dependence may or may not have desirable statistical properties such as being unbiased, or asymptotically consistent, based on the spatial structure of the population from which the data were sampled.

Sensitivity to the data distribution can be used to an advantage. For example, scaled correlation is designed to use the sensitivity to the range in order to pick out correlations between fast components of time series.[14] By reducing the range of values in a controlled manner, the correlations on long time scale are filtered out and only the correlations on short time scales are revealed.

7.5   Correlation matrices

The correlation matrix of n random variables $X_1, ..., X_n$ is the $n \times n$ matrix whose i,j entry is corr(X_i, X_j). If the measures of correlation used are product-moment coefficients, the correlation matrix is the same as the covariance matrix of the standardized random variables $X_i / \sigma(X_i)$ for $i = 1, ..., n$. This applies to both the matrix of population correlations (in which case "σ" is the population standard deviation), and to the matrix of sample correlations (in which case "σ" denotes the sample standard deviation). Consequently, each is necessarily a positive-semidefinite matrix.

The correlation matrix is symmetric because the correlation between X_i and X_j is the same as the correlation between X_j and X_i.

7.6   Common misconceptions

7.6.1   Correlation and causality

Main article: Correlation does not imply causation
See also: Normally distributed and uncorrelated does not imply independent

The conventional dictum that "correlation does not imply causation" means that correlation cannot be used to infer a causal relationship between the variables.[15] This dictum should not be taken to mean that correlations cannot indicate the potential existence of causal relations. However, the causes underlying the correlation, if any, may be indirect and unknown, and high correlations also overlap with identity relations (tautologies), where no causal process exists. Consequently, establishing a correlation between two variables is not a sufficient condition to establish a causal relationship (in either direction).

A correlation between age and height in children is fairly causally transparent, but a correlation between mood and health in people is less so. Does improved mood lead to improved health, or does good health lead to good mood, or both? Or does some other factor underlie both? In other words, a correlation can be taken as evidence for a possible causal relationship, but cannot indicate what the causal relationship, if any, might be.

7.6.2   Correlation and linearity

The Pearson correlation coefficient indicates the strength of a *linear* relationship between two variables, but its value generally does not completely characterize their relationship.[16] In particular, if the conditional mean of Y given X, denoted E($Y|X$), is not linear in X, the correlation coefficient will not fully determine the form of E($Y|X$).

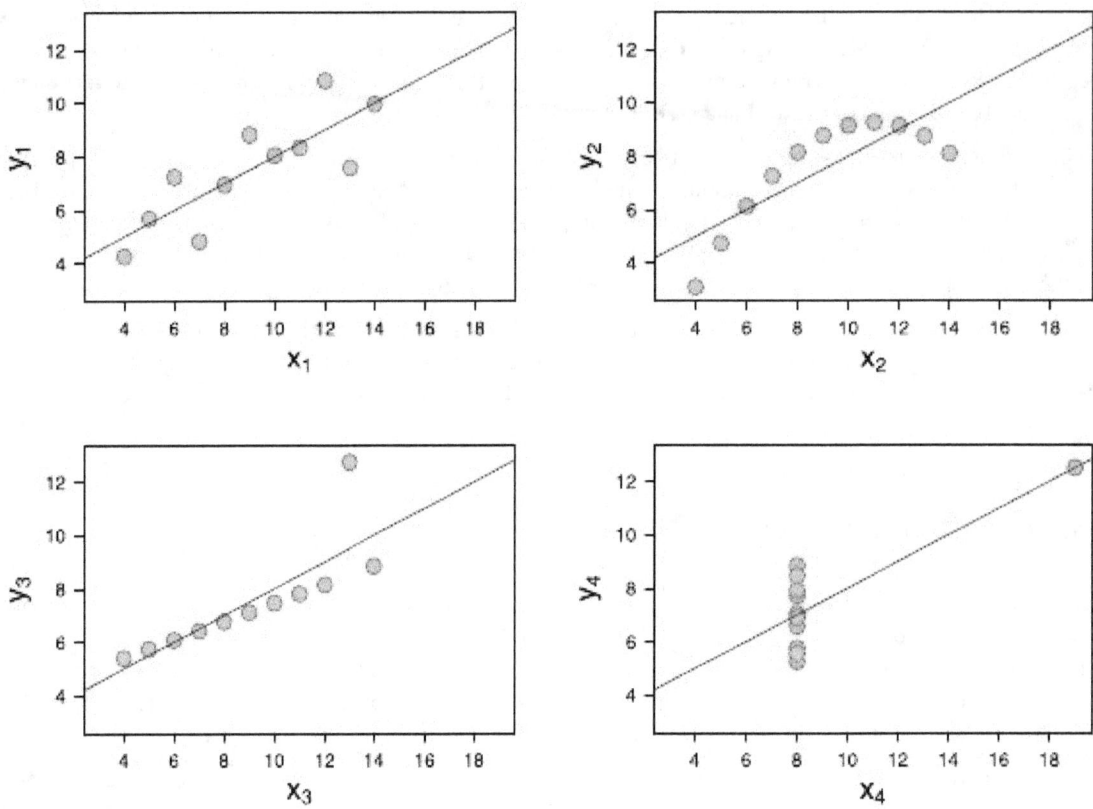

Four sets of data with the same correlation of 0.816

The image on the right shows scatterplots of Anscombe's quartet, a set of four different pairs of variables created by Francis Anscombe.[17] The four y variables have the same mean (7.5), variance (4.12), correlation (0.816) and regression line ($y = 3 + 0.5x$). However, as can be seen on the plots, the distribution of the variables is very different. The first one (top left) seems to be distributed normally, and corresponds to what one would expect when considering two variables correlated and following the assumption of normality. The second one (top right) is not distributed normally; while an obvious relationship between the two variables can be observed, it is not linear. In this case the Pearson correlation coefficient does not indicate that there is an exact functional relationship: only the extent to which that relationship can be approximated by a linear relationship. In the third case (bottom left), the linear relationship is perfect, except for one outlier which exerts enough influence to lower the correlation coefficient from 1 to 0.816. Finally, the fourth example (bottom right) shows another example when one outlier is enough to produce a high correlation coefficient, even though the relationship between the two variables is not linear.

These examples indicate that the correlation coefficient, as a summary statistic, cannot replace visual examination of the data. Note that the examples are sometimes said to demonstrate that the Pearson correlation assumes that the data follow a normal distribution, but this is not correct.[4]

7.7 Bivariate normal distribution

If a pair (X, Y) of random variables follows a bivariate normal distribution, the conditional mean $E(X|Y)$ is a linear function of Y, and the conditional mean $E(Y|X)$ is a linear function of X. The correlation coefficient r between X and Y, along with the marginal means and variances of X and Y, determines this linear relationship:

$$E(Y \mid X) = E(Y) + r\sigma_y \frac{X - E(X)}{\sigma_x}.$$

where $E(X)$ and $E(Y)$ are the expected values of X and Y, respectively, and σx and σy are the standard deviations of X and Y, respectively.

7.8 Partial correlation

Main article: Partial correlation

If a population or data-set is characterized by more than two variables, a partial correlation coefficient measures the strength of dependence between a pair of variables that is not accounted for by the way in which they both change in response to variations in a selected subset of the other variables.

7.9 See also

- Association (statistics)

- Autocorrelation

- Canonical correlation

- Coefficient of determination

- Cointegration

- Concordance correlation coefficient

- Cophenetic correlation

- Copula

- Correlation function

- Covariance and correlation

- Cross-correlation

- Ecological correlation

- Fraction of variance unexplained

- Genetic correlation

- Goodman and Kruskal's lambda

- Illusory correlation

- Interclass correlation

- Intraclass correlation

- Lift (data mining)

- Modifiable areal unit problem

- Multiple correlation

- Point-biserial correlation coefficient

- Quadrant count ratio

- Spurious correlation

- Statistical arbitrage

- Subindependence

7.10   References

[1]   Croxton, Frederick Emory; Cowden, Dudley Johnstone; Klein, Sidney (1968) *Applied General Statistics*, Pitman. ISBN 9780273403159 (page 625)

[2]   Dietrich, Cornelius Frank (1991) *Uncertainty, Calibration and Probability: The Statistics of Scientific and Industrial Measurement* 2nd Edition, A. Higler. ISBN 9780750300605 (Page 331)

[3]   Aitken, Alexander Craig (1957) *Statistical Mathematics* 8th Edition. Oliver & Boyd. ISBN 9780050013007 (Page 95)

[4]   Rodgers, J. L.; Nicewander, W. A. (1988). "Thirteen ways to look at the correlation coefficient". *The American Statistician* **42** (1): 59–66. doi:10.1080/00031305.1988.10475524. JSTOR 2685263.

[5]   Dowdy, S. and Wearden, S. (1983). "Statistics for Research", Wiley. ISBN 0-471-08602-9 pp 230

[6]   Francis, DP; Coats AJ; Gibson D (1999). "How high can a correlation coefficient be?". *Int J Cardiol* **69** (2): 185–199. doi:10.1016/S0167-5273(99)00028-5.

[7]   Yule, G.U and Kendall, M.G. (1950). "An Introduction to the Theory of Statistics", 14th Edition (5th Impression 1968). Charles Griffin & Co. pp 258–270

[8]   Kendall, M. G. (1955) "Rank Correlation Methods", Charles Griffin & Co.

[9]   Mahdavi Damghani B. (2013). "The Non-Misleading Value of Inferred Correlation: An Introduction to the Cointelation Model". *Wilmott Magazine*. doi:10.1002/wilm.10252.

[10] Székely, G. J. Rizzo; Bakirov, N. K. (2007). "Measuring and testing independence by correlation of distances". *Annals of Statistics* **35** (6): 2769–2794. doi:10.1214/009053607000000505.

[11] Székely, G. J.; Rizzo, M. L. (2009). "Brownian distance covariance". *Annals of Applied Statistics* **3** (4): 1233–1303. doi:10.1214/09-AOAS312.

[12] Lopez-Paz D. and Hennig P. and Schölkopf B. (2013). "The Randomized Dependence Coefficient", "Conference on Neural Information Processing Systems" Reprint

[13] Thorndike, Robert Ladd (1947). *Research problems and techniques (Report No. 3)*. Washington DC: US Govt. print. off.

[14] Nikolić, D; Muresan, RC; Feng, W; Singer, W (2012). "Scaled correlation analysis: a better way to compute a cross-correlogram". *European Journal of Neuroscience*: 1–21. doi:10.1111/j.1460-9568.2011.07987.x.

[15] Aldrich, John (1995). "Correlations Genuine and Spurious in Pearson and Yule". *Statistical Science* **10**(4): 364–376.doi:10.1214/ JSTOR 2246135.

[16] Mahdavi Damghani, Babak (2012). "The Misleading Value of Measured Correlation". *Wilmott* **2012**(1): 64–73.doi:10.1002/

[17] Anscombe, Francis J. (1973). "Graphs in statistical analysis". *The American Statistician* **27**: 17–21. doi:10.2307/2682899. JSTOR 2682899.

7.11   Further reading

- Cohen, J., Cohen P., West, S.G., & Aiken, L.S. (2002). *Applied multiple regression/correlation analysis for the behavioral sciences (3rd ed.)*. Psychology Press. ISBN 0-8058-2223-2.

- Hazewinkel, Michiel, ed. (2001), "Correlation (in statistics)", *Encyclopedia of Mathematics*, Springer, ISBN 978-1-55608-010-4

- Oestreicher, J. & D. R. (February 26, 2015). *Plague of Equals: A science thriller of international disease, politics and drug discovery*. California: Omega Cat Press. p. 408. ISBN 978-0963175540.

7.12   External links

- MathWorld page on the (cross-)correlation coefficient/s of a sample

- Compute significance between two correlations, for the comparison of two correlation values.

- A MATLAB Toolbox for computing Weighted Correlation Coefficients

- Proof that the Sample Bivariate Correlation Coefficient has Limits ±1

- Interactive Flash simulation on the correlation of two normally distributed variables by Juha Puranen.

- Correlation analysis. Biomedical Statistics

- R-Psychologist Correlation visualization of correlation between two numeric variables

Chapter 8

Dependent and independent variables

Variable used in an experiment or modelling can be divided into three types: "**dependent variable**", "**independent variable**", or other. The "dependent variable" represents the output or effect, or is tested to see if it is the effect. The "independent variable" represents the input or cause, or is tested to see if it is the cause. Other variables may also be observed for various reasons.

8.1 Use

8.1.1 Mathematics

In mathematics, a function is a rule for taking an input (usually a number or set of numbers)[2] and providing an output (which is also usually a number).[2] A symbol that stands for an arbitrary input is called an **independent variable**, while a symbol that stands for an arbitrary output is called a **dependent variable**.[3] The most common symbol for the input is x, and the most common symbol for the output is y; the function itself is commonly written $y = f(x)$.[3][4]

It is possible to have multiple independent variables and/or multiple dependent variables. For instance, in multivariable calculus, one often encounters functions of the form $z = f(x, y)$, where z is a dependent variable and x and y are independent variables.[5] Functions with multiple outputs are often written as vector-valued functions.

In advanced mathematics, a function between a set **X** and a set **Y** is a subset of the Cartesian product $X \times Y$ such that every element of **X** appears in an ordered pair with exactly one element of **Y**. In this situation, a symbol representing an element of **X** may be called a dependent variable and a symbol representing an element of **Y** may be called an independent variable, such as when **X** is a manifold and the symbol x represents an arbitrary point in the manifold.[6] However, many advanced textbooks do not distinguish between dependent and independent variables.[7]

8.1.2 Statistics

In a statistics experiment, the dependent variable is the event studied and expected to change whenever the independent variable is altered.[8]

In data mining tools (for multivariate statistics and machine learning), the depending variable is assigned a *role* as **target variable** (or in some tools as *label attribute*), while a dependent variable may be assigned a role as *regular variable*.[9] Known values for the target variable are provided for the training data set and test data set, but should be predicted for other data. The target variable is used in supervised learning algorithms but not in non-supervised learning.

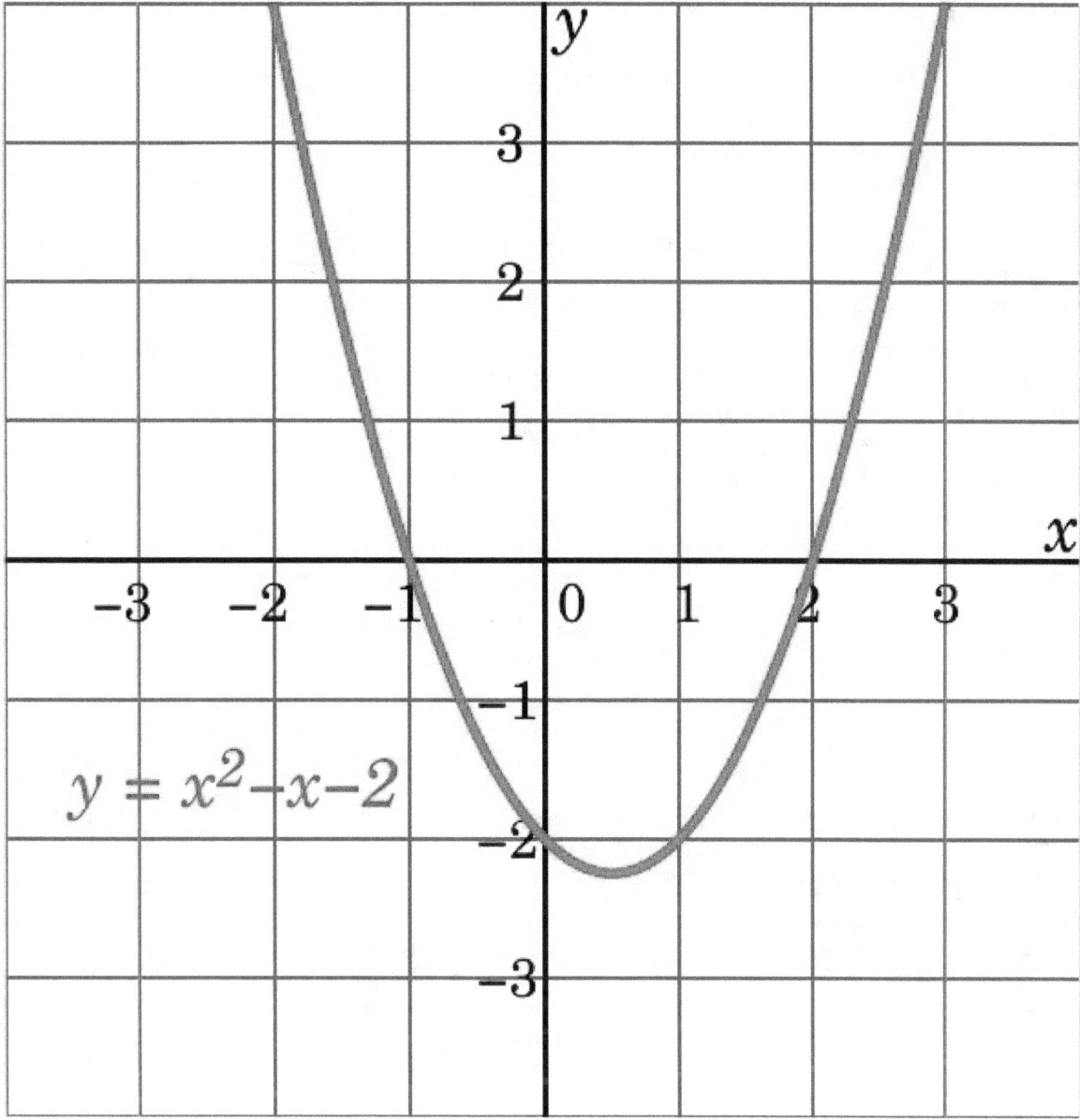

In calculus, a function is typically graphed with the horizontal axis representing the independent variable and the vertical axis representing the dependent variable.[1] In this function, y is the dependent variable and x is the independent variable.

8.1.3   Modelling

In mathematical modelling, the dependent variable is studied to see if and how much it varies as the independent variables vary. In the simple stochastic linear model $y_i = a + bx_i + e_i$ the term y_i is the ith value of the dependent variable and x_i is i^{th} value of the independent variable. The term e_i is known as the "error" and contains the variability of the dependent variable not explained by the independent variable.

With multiple independent variables, the expression is: $y_i = a + bx_1 + bx_2 + ... + bx_n + e_i$, where n is the number of independent variables.

8.1.4 Simulation

In simulation, the dependent variable is changed in response to changes in the independent variables.

8.2 Statistics synonyms

An independent variable is also known as a "predictor variable", "regressor", "controlled variable", "manipulated variable", "explanatory variable", "exposure variable" (see reliability theory), "risk factor" (see medical statistics), "feature" (in machine learning and pattern recognition) or an "input variable."[10][11]

A dependent variable is also known as a "response variable", "regressand", "predicated variable", "measured variable", "explained variable", "experimental variable", "responding variable", "outcome variable", and "output variable".[11]

"Explanatory variable" is preferred by some authors over "independent variable" when the quantities treated as "independent variables" may not be statistically independent.[12][13] If the independent variable is referred to as an "explanatory variable" then the term "response variable" is preferred by some authors for the dependent variable.[11][12][13]

"Explained variable" is preferred by some authors over "dependent variable" when the quantities treated as "dependent variables" may not be statistically dependent.[14] If the dependent variable is referred to as an "explained variable" then the term "predictor variable" is preferred by some authors for the independent variable.[14]

Variables may also be referred to by their form: continuous, binary/dichotomous, nominal categorical, and ordinal categorical, among others.

8.3 Other variables

A variable may be thought to alter the dependent or independent variables, but may not actually be the focus of the experiment. So that variable will be kept constant or monitored to try to minimise its effect on the experiment. Such variables may be designated as either a "controlled variable" , "control variable", or "extraneous variable".

Extraneous variables, if included in a regression as independent variables, may aid a researcher with accurate response parameter estimation, prediction, and goodness of fit, but are not of substantive interest to the hypothesis under examination. For example, in a study examining the effect of post-secondary education on lifetime earnings, some extraneous variables might be gender, ethnicity, social class, genetics, intelligence, age, and so forth. A variable is extraneous only when it can be assumed (or shown) to influence the dependent variable. If included in a regression, it can improve the fit of the model. If it is excluded from the regression and if it has a non-zero covariance with one or more of the independent variables of interest, its omission will bias the regression's result for the effect of that independent variable of interest. This effect is called confounding or omitted variable bias; in these situations, design changes and/or statistical control is necessary.

Extraneous variables are often classified into three types:

1. Subject variables, which are the characteristics of the individuals being studied that might affect their actions. These variables include age, gender, health status, mood, background, etc.

2. Blocking variables or experimental variables are characteristics of the persons conducting the experiment which might influence how a person behaves. Gender, the presence of racial discrimination, language, or other factors may qualify as such variables.

3. Situational variables are features of the environment in which the study or research was conducted, which have a bearing on the outcome of the experiment in a negative way. Included are the air temperature, level of activity, lighting, and the time of day.

In quasi-experiments, differentiating between dependent and other variables may be downplayed in favour of differentiating between those variables that can be altered by the researcher and those that cannot. Variables in quasi-experiments may be referred to as "extraneous variables", "subject variables", "blocking variables", "situational variables", "pseudo-independent variables", "ex post facto variables", "natural group variables" or "non-manipulated variables".

In modelling, variability that is not covered by the independent variable is designated by e_i and is known as the "residual", "side effect", "error", "unexplained share", "residual variable", or "tolerance".

8.4   Examples

- Effects of vitamin C on life span

  In a study whether taking vitamin C pills daily make people live longer, researchers will dictate the vitamin C intake of a group of people over time. One part of the group will be given vitamin C pills daily. The other part of the group will be given a placebo pill. Nobody in the group knows which part they are in. The researchers will check the life span of the people in both groups. Here, the dependent variable is the life span and the independent variable is a binary variable for the use or non-use of vitamin C.

- Effect of fertilizer on plant growth

  In a study measuring the influence of different quantities of fertilizer on plant growth, the independent variable would be the amount of fertilizer used. The dependent variable would be the growth in height or mass of the plant. The controlled variables would be the type of plant, the type of fertilizer, the amount of sunlight the plant gets, the size of the pots, etc.

- Effect of drug dosage on symptom severity

  In a study of how different doses of a drug affect the severity of symptoms, a researcher could compare the frequency and intensity of symptoms when different doses are administered. Here the independent variable is the dose and the dependent variable is the frequency/intensity of symptoms.

- Effect of temperature on pigmentation

  In measuring the amount of color removed from beetroot samples at different temperatures, temperature is the independent variable and amount of pigment removed is the dependent variable.

- Effect of education on wealth

  In sociology, in measuring the effect of education on income or wealth, the dependent variable is level of income/wealth and the independent variable is the education level of the individual.

8.5   See also

- Abscissa

- Ordinate

- Blocking (statistics)

8.6 References

[1] Hastings, Nancy Baxter. Workshop calculus: guided exploration with review. Vol. 2. Springer Science & Business Media, 1998. p. 31

[2] Carlson, Robert. A concrete introduction to real analysis. CRC Press, 2006. p.183

[3] Stewart, James. Calculus. Cengage Learning, 2011. Section 1.1

[4] Anton, Howard, Irl C. Bivens, and Stephen Davis. Calculus Single Variable. John Wiley & Sons, 2012. Section 0.1

[5] Larson, Ron, and Bruce Edwards. Calculus. Cengage Learning, 2009. Section 13.1

[6] Hrbacek, Karel, and Thomas Jech. Introduction to Set Theory, Revised and Expanded. Vol. 220. Crc Press, 1999. p. 26

[7] For instance, a Google Books search for "independent variable" on Mar 18, 2015 brought up 0 hits in the following advanced textbooks:

- Munkres, James R. Topology: a first course. Vol. 23. Englewood Cliffs, NJ: Prentice-Hall, 1975.
- Hungerford, Thomas. Abstract algebra: an introduction. Cengage Learning, 2012.
- Abbott, Stephen. Understanding analysis. Springer Science & Business Media, 2010.

[8] *Random House Webster's Unabridged Dictionary*. Random House, Inc. 2001. Page 534, 971. ISBN 0-375-42566-7.

[9] English Manual version 1.0 for RapidMiner 5.0, October 2013.

[10] Dodge, Y. (2003) *The Oxford Dictionary of Statistical Terms*, OUP. ISBN 0-19-920613-9 (entry for "independent variable")

[11] Dodge, Y. (2003) *The Oxford Dictionary of Statistical Terms*, OUP. ISBN 0-19-920613-9 (entry for "regression")

[12] Everitt, B.S. (2002) Cambridge Dictionary of Statistics, CUP. ISBN 0-521-81099-X

[13] Dodge, Y. (2003) *The Oxford Dictionary of Statistical Terms*, OUP. ISBN 0-19-920613-9

[14] Ash Narayan Sah (2009) Data Analysis Using Microsoft Excel, New Delhi. ISBN 978-81-7446-716-4

Chapter 9

Cartesian coordinate system

A **Cartesian coordinate system** is a coordinate system that specifies each point uniquely in a plane by a pair of numerical **coordinates**, which are the signed distances from the point to two fixed perpendicular directed lines, measured in the same unit of length. Each reference line is called a *coordinate axis* or just *axis* of the system, and the point where they meet is its *origin*, usually at ordered pair $(0, 0)$. The coordinates can also be defined as the positions of the perpendicular projections of the point onto the two axes, expressed as signed distances from the origin.

One can use the same principle to specify the position of any point in three-dimensional space by three Cartesian coordinates, its signed distances to three mutually perpendicular planes (or, equivalently, by its perpendicular projection onto three mutually perpendicular lines). In general, n Cartesian coordinates (an element of real n-space) specify the point in an n-dimensional Euclidean space for any dimension n. These coordinates are equal, up to sign, to distances from the point to n mutually perpendicular hyperplanes.

The invention of Cartesian coordinates in the 17th century by René Descartes (Latinized name: *Cartesius*) revolutionized mathematics by providing the first systematic link between Euclidean geometry and algebra. Using the Cartesian coordinate system, geometric shapes (such as curves) can be described by **Cartesian equations**: algebraic equations involving the coordinates of the points lying on the shape. For example, a circle of radius 2 in a plane may be described as the set of all points whose coordinates x and y satisfy the equation $x^2 + y^2 = 4$.

Cartesian coordinates are the foundation of analytic geometry, and provide enlightening geometric interpretations for many other branches of mathematics, such as linear algebra, complex analysis, differential geometry, multivariate calculus, group theory and more. A familiar example is the concept of the graph of a function. Cartesian coordinates are also essential tools for most applied disciplines that deal with geometry, including astronomy, physics, engineering and many more. They are the most common coordinate system used in computer graphics, computer-aided geometric design and other geometry-related data processing.

9.1 History

The adjective *Cartesian* refers to the French mathematician and philosopher René Descartes (who used the name *Cartesius* in Latin).

The idea of this system was developed in 1637 in writings by Descartes and independently by Pierre de Fermat, although Fermat also worked in three dimensions and did not publish the discovery.[1] Both authors used a single axis in their treatments and have a variable length measured in reference to this axis. The concept of using a pair of axes was introduced later, after Descartes' *La Géométrie* was translated into Latin in 1649 by Frans van Schooten and his students. These commentators introduced several concepts while trying to clarify the ideas contained in Descartes' work.[2]

The development of the Cartesian coordinate system would play a fundamental role in the development of the calculus by Isaac Newton and Gottfried Wilhelm Leibniz.[3]

Nicole Oresme, a French cleric and friend of the Dauphin (later to become King Charles V) of the 14th Century, used

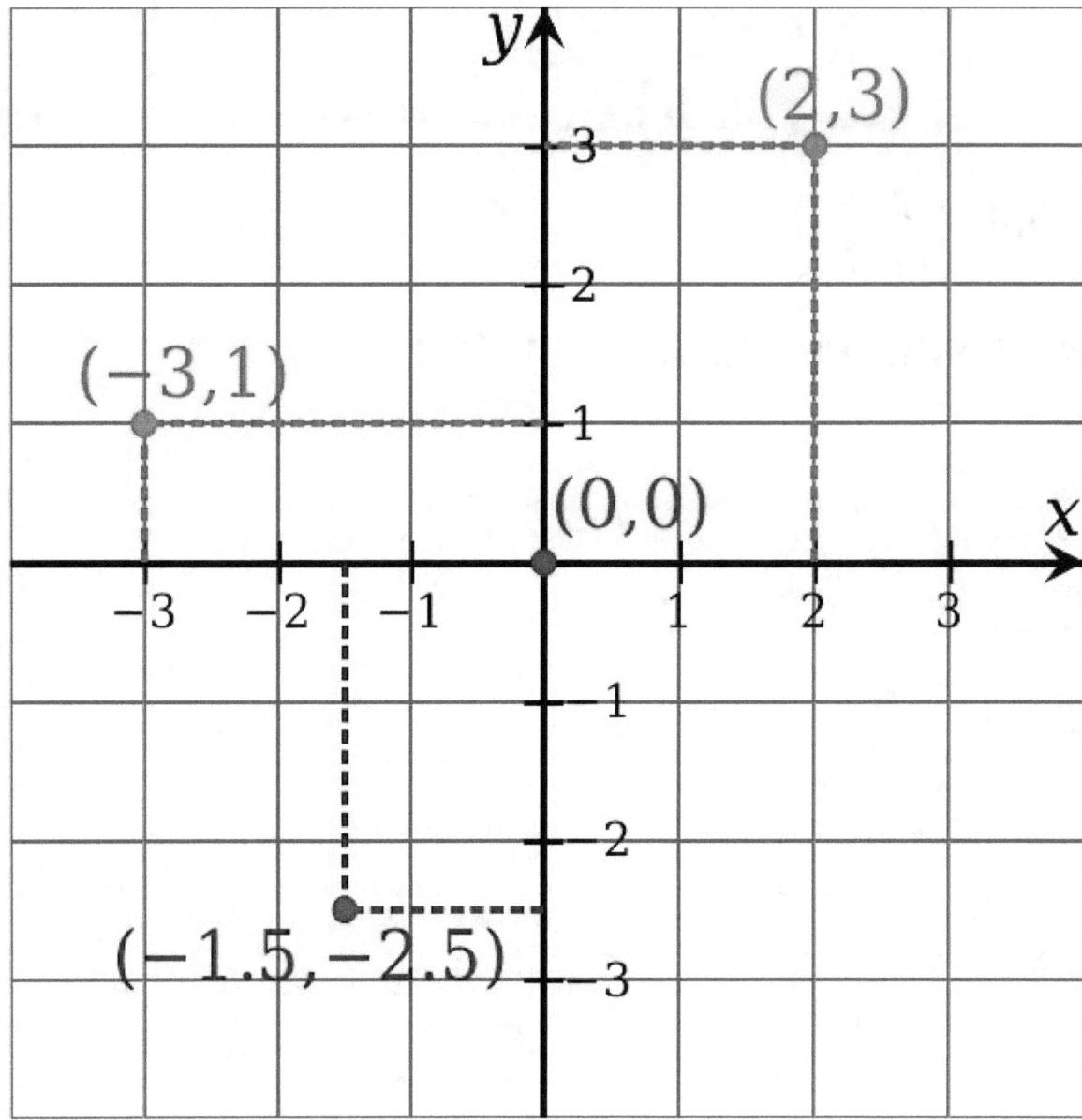

Illustration of a Cartesian coordinate plane. Four points are marked and labeled with their coordinates: (2,3) in green, (−3,1) in red, (−1.5,−2.5) in blue, and the origin (0,0) in purple.

constructions similar to Cartesian coordinates well before the time of Descartes and Fermat.

Many other coordinate systems have been developed since Descartes, such as the polar coordinates for the plane, and the spherical and cylindrical coordinates for three-dimensional space.

9.2 Description

9.2.1 One dimension

Main article: Number line

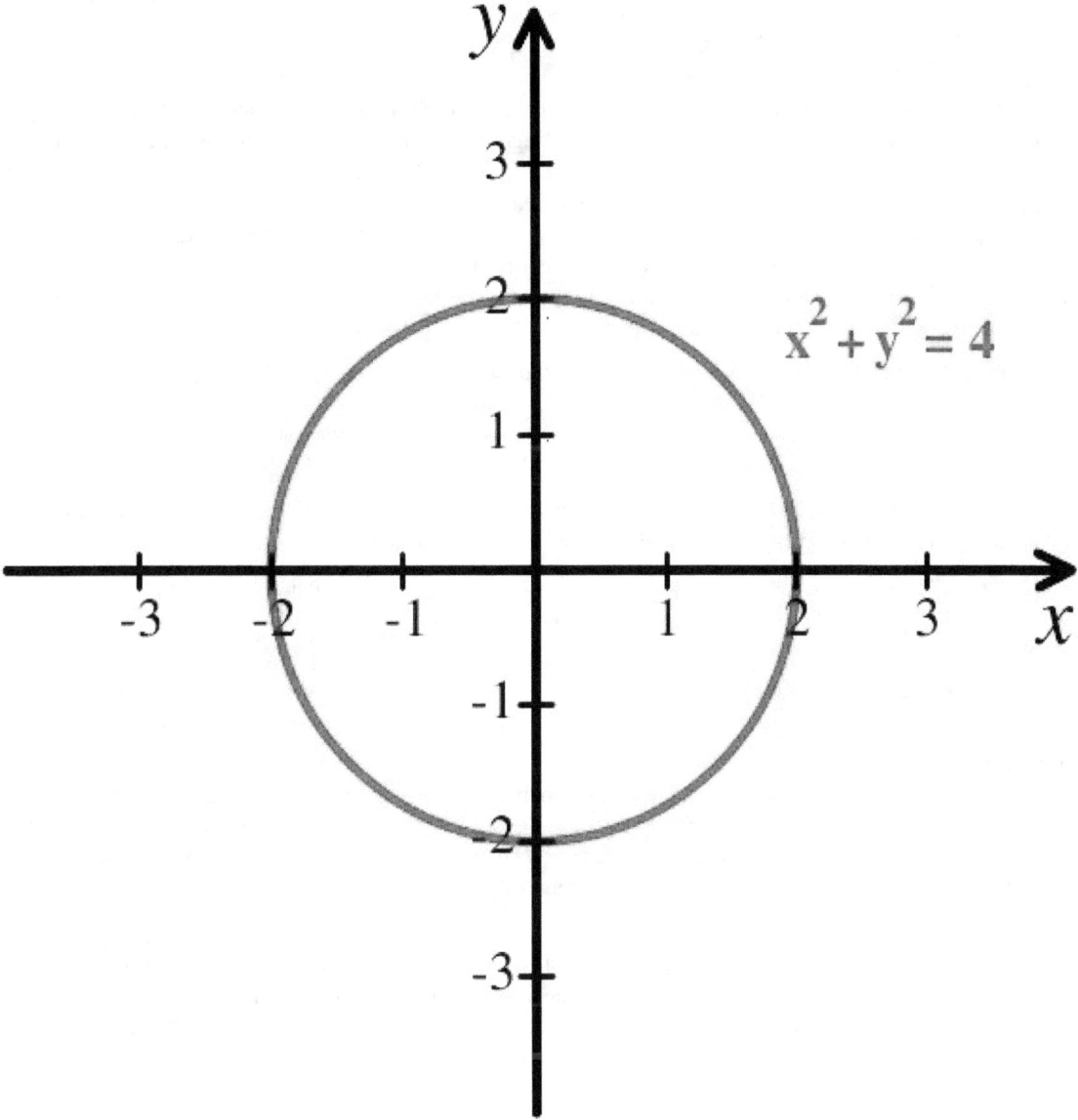

Cartesian coordinate system with a circle of radius 2 centered at the origin marked in red. The equation of a circle is $(x - a)^2 + (y - b)^2 = r^2$ where a and b are the coordinates of the center (a, b) and r is the radius.

Choosing a Cartesian coordinate system for a one-dimensional space—that is, for a straight line—involves choosing a point O of the line (the origin), a unit of length, and an orientation for the line. An orientation chooses which of the two half-lines determined by O is the positive, and which is negative; we then say that the line "is oriented" (or "points") from the negative half towards the positive half. Then each point P of the line can be specified by its distance from O, taken with a + or − sign depending on which half-line contains P.

A line with a chosen Cartesian system is called a **number line**. Every real number has a unique location on the line. Conversely, every point on the line can be interpreted as a number in an ordered continuum such as the real numbers.

9.2.2   Two dimensions

Further information: Two-dimensional space

The Cartesian coordinate system in two dimensions (also called a **rectangular coordinate system**) is defined by an ordered pair of perpendicular lines (axes), a single unit of length for both axes, and an orientation for each axis. (Early systems allowed "oblique" axes, that is, axes that did not meet at right angles.) The lines are commonly referred to as the x- and y-axes where the x-axis is taken to be horizontal and the y-axis is taken to be vertical. The point where the axes meet is taken as the origin for both, thus turning each axis into a number line. For a given point P, a line is drawn through P perpendicular to the x-axis to meet it at X and second line is drawn through P perpendicular to the y-axis to meet it at Y. The coordinates of P are then X and Y interpreted as numbers x and y on the corresponding number lines. The coordinates are written as an ordered pair (x, y).

The point where the axes meet is the common origin of the two number lines and is simply called the *origin*. It is often labeled O and if so then the axes are called Ox and Oy. A plane with x- and y-axes defined is often referred to as the Cartesian plane or xy-plane. The value of x is called the x-coordinate or **abscissa** and the value of y is called the y-coordinate or **ordinate**.

The choices of letters come from the original convention, which is to use the latter part of the alphabet to indicate unknown values. The first part of the alphabet was used to designate known values.

In the Cartesian plane, reference is sometimes made to a unit circle or a unit hyperbola.

9.2.3   Three dimensions

Further information: Three-dimensional space
 Choosing a Cartesian coordinate system for a three-dimensional space means choosing an ordered triplet of lines (axes) that are pair-wise perpendicular, have a single unit of length for all three axes and have an orientation for each axis. As in the two-dimensional case, each axis becomes a number line. The coordinates of a point P are obtained by drawing a line through P perpendicular to each coordinate axis, and reading the points where these lines meet the axes as three numbers of these number lines.

Alternatively, the coordinates of a point P can also be taken as the (signed) distances from P to the three planes defined by the three axes. If the axes are named x, y, and z, then the x-coordinate is the distance from the plane defined by the y and z axes. The distance is to be taken with the $+$ or $-$ sign, depending on which of the two half-spaces separated by that plane contains P. The y and z coordinates can be obtained in the same way from the xz- and xy-planes respectively.

9.2.4   Higher dimensions

A Euclidean plane with a chosen Cartesian system is called a **Cartesian plane**. Since Cartesian coordinates are unique and non-ambiguous, the points of a Cartesian plane can be identified with pairs of real numbers; that is with the Cartesian product $\mathbb{R}^2 = \mathbb{R} \times \mathbb{R}$, where $\mathbb{R}$ is the set of all reals. In the same way, the points any Euclidean space of dimension n be identified with the tuples (lists) of n real numbers, that is, with the Cartesian product $\mathbb{R}^n$.

9.2.5   Generalizations

The concept of Cartesian coordinates generalizes to allow axes that are not perpendicular to each other, and/or different units along each axis. In that case, each coordinate is obtained by projecting the point onto one axis along a direction that is parallel to the other axis (or, in general, to the hyperplane defined by all the other axes). In such an **oblique coordinate system** the computations of distances and angles must be modified from that in standard Cartesian systems, and many standard formulas (such as the Pythagorean formula for the distance) do not hold.

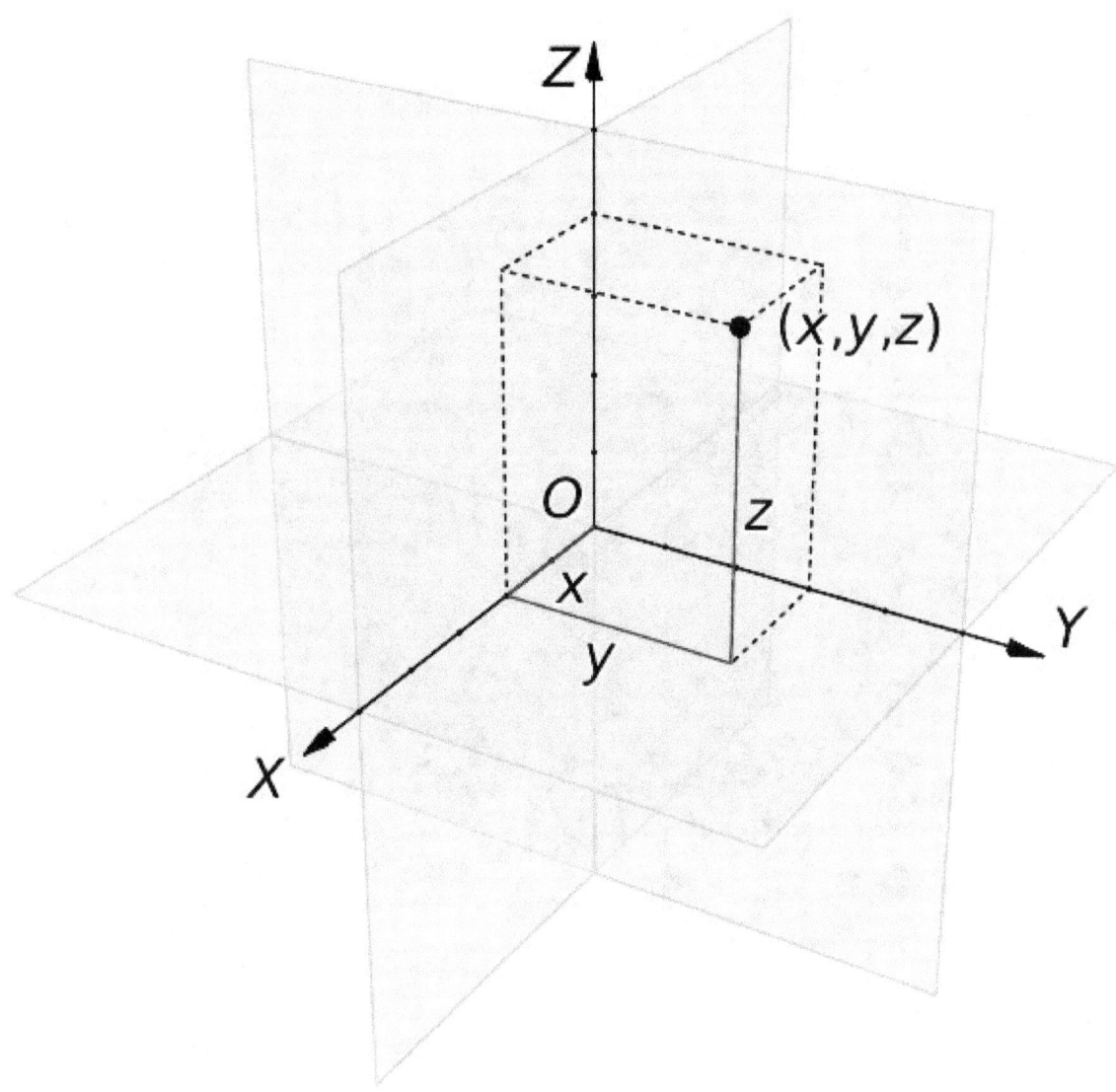

A three dimensional Cartesian coordinate system, with origin O and axis lines X, Y and Z, oriented as shown by the arrows. The tick marks on the axes are one length unit apart. The black dot shows the point with coordinates x = 2, y = 3, and z = 4, or (2,3,4).

9.3   Notations and conventions

The Cartesian coordinates of a point are usually written in parentheses and separated by commas, as in (10, 5) or (3, 5, 7). The origin is often labelled with the capital letter O. In analytic geometry, unknown or generic coordinates are often denoted by the letters x and y on the plane, and x, y, and z in three-dimensional space. This custom comes from a convention of algebra, which use letters near the end of the alphabet for unknown values (such as were the coordinates of points in many geometric problems), and letters near the beginning for given quantities.

These conventional names are often used in other domains, such as physics and engineering, although other letters may be used. For example, in a graph showing how a pressure varies with time, the graph coordinates may be denoted t and p. Each axis is usually named after the coordinate which is measured along it; so one says the *x-axis*, the *y-axis*, the *t-axis*, etc.

Another common convention for coordinate naming is to use subscripts, as in x_1, x_2, ... xn for the n coordinates in an

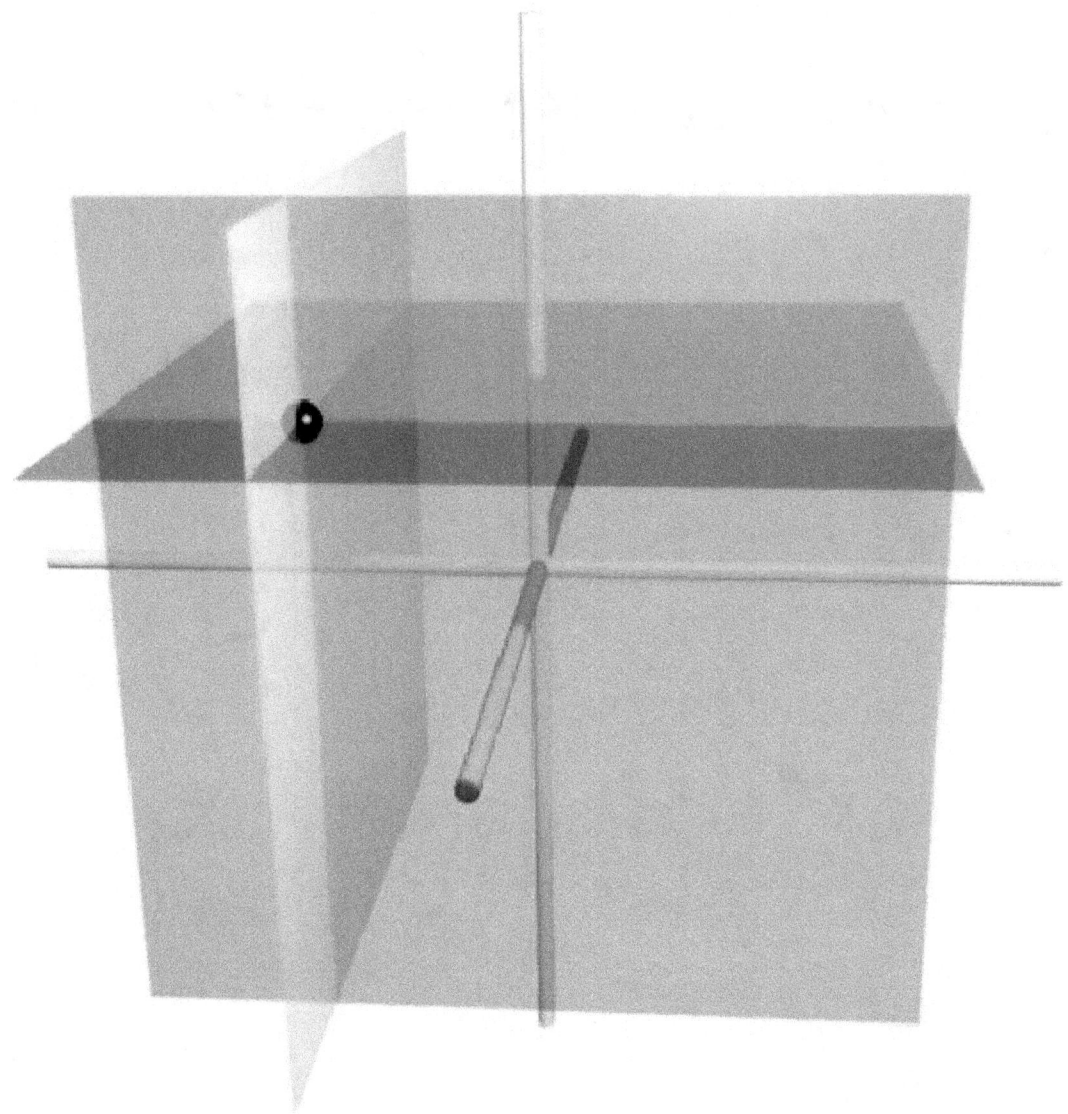

The coordinate surfaces of the Cartesian coordinates (x, y, z). The z-axis is vertical and the x-axis is highlighted in green. Thus, the red plane shows the points with x = 1, the blue plane shows the points with z = 1, and the yellow plane shows the points with y = −1. The three surfaces intersect at the point P (shown as a black sphere) with the Cartesian coordinates (1, −1, 1).

n-dimensional space; especially when n is greater than 3, or not specified. Some authors prefer the numbering $x_0, x_1, \ldots x_{n-1}$. These notations are especially advantageous in computer programming: by storing the coordinates of a point as an array, instead of a record, the subscript can serve to index the coordinates.

In mathematical illustrations of two-dimensional Cartesian systems, the first coordinate (traditionally called the abscissa) is measured along a horizontal axis, oriented from left to right. The second coordinate (the ordinate) is then measured along a vertical axis, usually oriented from bottom to top.

However, computer graphics and image processing often use a coordinate system with the y axis oriented downwards on the computer display. This convention developed in the 1960s (or earlier) from the way that images were originally stored in display buffers.

For three-dimensional systems, a convention is to portray the xy-plane horizontally, with the z axis added to represent height (positive up). Furthermore, there is a convention to orient the x-axis toward the viewer, biased either to the right or left. If a diagram (3D projection or 2D perspective drawing) shows the x and y axis horizontally and vertically, respectively, then the z axis should be shown pointing "out of the page" towards the viewer or camera. In such a 2D diagram of a 3D coordinate system, the z axis would appear as a line or ray pointing down and to the left or down and to the right, depending on the presumed viewer or camera perspective. In any diagram or display, the orientation of the three axes, as a whole, is arbitrary. However, the orientation of the axes relative to each other should always comply with the right-hand rule, unless specifically stated otherwise. All laws of physics and math assume this right-handedness, which ensures consistency.

For 3D diagrams, the names "abscissa" and "ordinate" are rarely used for x and y, respectively. When they are, the z-coordinate is sometimes called the **applicate**. The words *abscissa*, *ordinate* and *applicate* are sometimes used to refer to coordinate axes rather than the coordinate values.[4]

9.3.1   Quadrants and octants

Main articles: Octant (solid geometry) and Quadrant (plane geometry)
  The axes of a two-dimensional Cartesian system divide the plane into four infinite regions, called **quadrants**, each

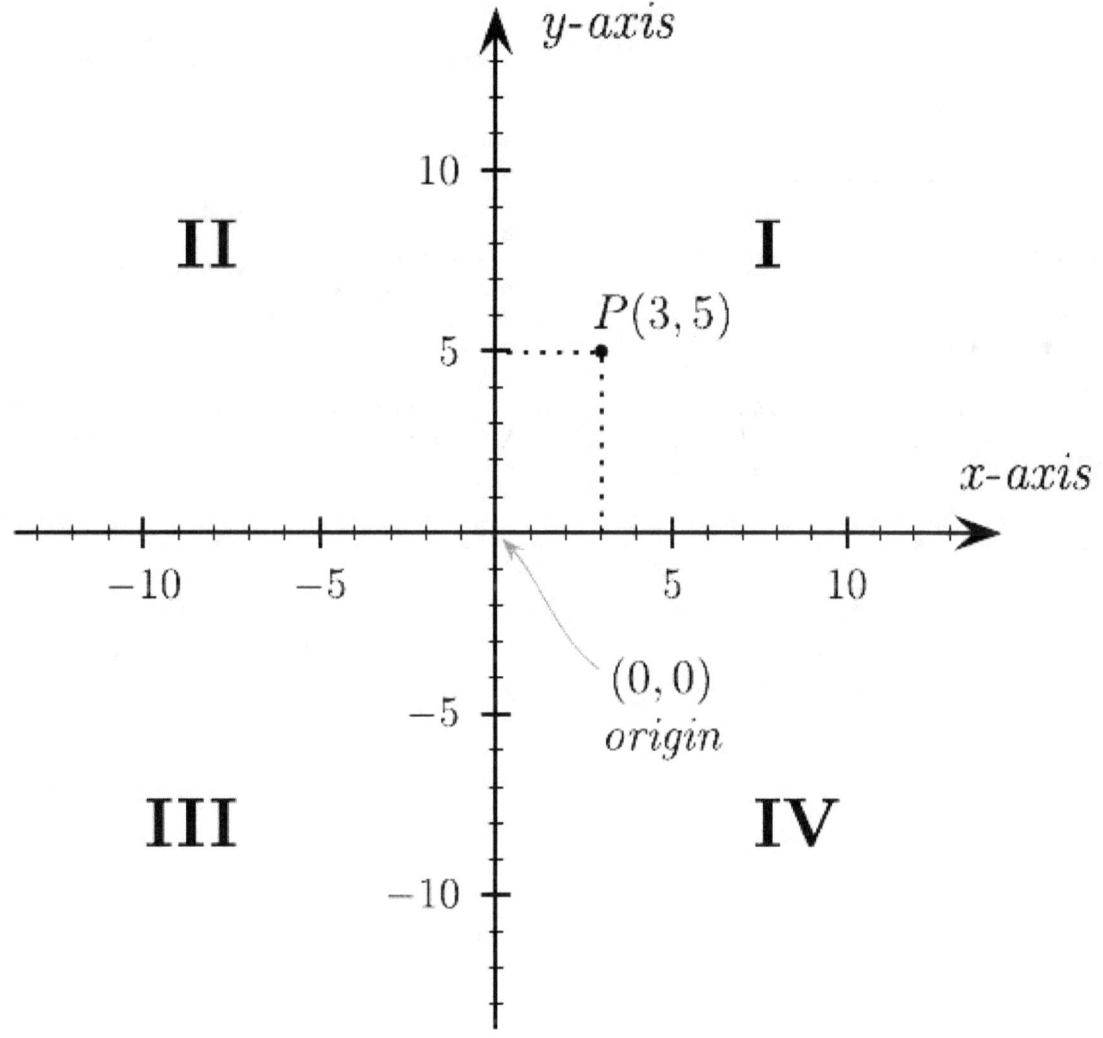

The four quadrants of a Cartesian coordinate system.

bounded by two half-axes. These are often numbered from 1st to 4th and denoted by Roman numerals: I (where the signs of the two coordinates are +,+), II (−,+), III (−,−), and IV (+,−). When the axes are drawn according to the mathematical custom, the numbering goes counter-clockwise starting from the upper right ("north-east") quadrant.

Similarly, a three-dimensional Cartesian system defines a division of space into eight regions or **octants**, according to the signs of the coordinates of the points. The convention used for naming a specific octant is to list its signs, e.g. (+ + +) or (− + −). The generalization of the quadrant and octant to an arbitrary number of dimensions is the **orthant**, and a similar naming system applies.

9.4 Cartesian formulae for the plane

9.4.1 Distance between two points

The Euclidean distance between two points of the plane with Cartesian coordinates (x_1, y_1) and (x_2, y_2) is

$$d = \sqrt{(x_2 - x_1)^2 + (y_2 - y_1)^2}.$$

This is the Cartesian version of Pythagoras's theorem. In three-dimensional space, the distance between points (x_1, y_1, z_1) and (x_2, y_2, z_2) is

$$d = \sqrt{(x_2 - x_1)^2 + (y_2 - y_1)^2 + (z_2 - z_1)^2},$$

which can be obtained by two consecutive applications of Pythagoras' theorem.

9.4.2 Euclidean transformations

The Euclidean transformations or **Euclidean motions** are the (bijective) mappings of points of the Euclidean plane to themselves which preserve distances between points. There are four types of these mappings (also called isometries): translations, rotations, reflections and glide reflections.[5]

Translation

Translating a set of points of the plane, preserving the distances and directions between them, is equivalent to adding a fixed pair of numbers (a, b) to the Cartesian coordinates of every point in the set. That is, if the original coordinates of a point are (x, y), after the translation they will be

$$(x', y') = (x + a, y + b).$$

Rotation

To rotate a figure counterclockwise around the origin by some angle θ is equivalent to replacing every point with coordinates (x, y) by the point with coordinates (x', y'), where

$$x' = x \cos \theta - y \sin \theta$$

$$y' = x \sin \theta + y \cos \theta.$$

Thus: $(x', y') = ((x \cos \theta - y \sin \theta), (x \sin \theta + y \cos \theta)).$

Reflection

If (x, y) are the Cartesian coordinates of a point, then $(-x, y)$ are the coordinates of its reflection across the second coordinate axis (the Y-axis), as if that line were a mirror. Likewise, $(x, -y)$ are the coordinates of its reflection across the first coordinate axis (the X-axis). In more generality, reflection across a line through the origin making an angle θ with the x-axis, is equivalent to replacing every point with coordinates (x, y) by the point with coordinates (x', y'), where

$$x' = x \cos 2\theta + y \sin 2\theta$$

$$y' = x \sin 2\theta - y \cos 2\theta.$$

Thus: $(x', y') = ((x \cos 2\theta + y \sin 2\theta), (x \sin 2\theta - y \cos 2\theta)).$

Glide reflection

A glide reflection is the composition of a reflection across a line followed by a translation in the direction of that line. It can be seen that the order of these operations does not matter (the translation can come first, followed by the reflection).

General matrix form of the transformations

These Euclidean transformations of the plane can all be described in a uniform way by using matrices. The result (x', y') of applying a Euclidean transformation to a point (x, y) is given by the formula

$$(x', y') = (x, y)A + b$$

where A is a 2×2 orthogonal matrix and $b = (b_1, b_2)$ is an arbitrary ordered pair of numbers;[6] that is,

$$x' = xA_{11} + yA_{21} + b_1$$

$$y' = xA_{12} + yA_{22} + b_2,$$

where

$$A = \begin{pmatrix} A_{11} & A_{12} \\ A_{21} & A_{22} \end{pmatrix}.$$ [Note the use of row vectors for point coordinates and that the matrix is written on the right.]

To be *orthogonal*, the matrix A must have orthogonal rows with same Euclidean length of one, that is,

$$A_{11}A_{21} + A_{12}A_{22} = 0$$

and

$$A_{11}^2 + A_{12}^2 = A_{21}^2 + A_{22}^2 = 1.$$

This is equivalent to saying that A times its transpose must be the identity matrix. If these conditions do not hold, the formula describes a more general affine transformation of the plane provided that the determinant of A is not zero.

The formula defines a translation if and only if A is the identity matrix. The transformation is a rotation around some point if and only if A is a rotation matrix, meaning that

$A_{11}A_{22} - A_{21}A_{12} = 1.$

A reflection or glide reflection is obtained when,

$A_{11}A_{22} - A_{21}A_{12} = -1.$

Assuming that translation is not used transformations can be combined by simply multiplying the associated transformation matrices.

Affine transformation

Another way to represent coordinate transformations in Cartesian coordinates is through affine transformations. In affine transformations an extra dimension is added and all points are given a value of 1 for this extra dimension. The advantage of doing this is that point translations can be specified in the final column of matrix A. In this way, all of the euclidean transformations become transactable as matrix point multiplications. The affine transformation is given by:

$$\begin{pmatrix} A_{11} & A_{21} & b_1 \\ A_{12} & A_{22} & b_2 \\ 0 & 0 & 1 \end{pmatrix} \begin{pmatrix} x \\ y \\ 1 \end{pmatrix} = \begin{pmatrix} x' \\ y' \\ 1 \end{pmatrix}.$$ [Note the matrix A from above was transposed. The matrix is on the left and column vectors for point coordinates are used.]

Using affine transformations multiple different euclidean transformations including translation can be combined by simply multiplying the corresponding matrices.

Scaling

An example of an affine transformation which is not a Euclidean motion is given by scaling. To make a figure larger or smaller is equivalent to multiplying the Cartesian coordinates of every point by the same positive number m. If (x, y) are the coordinates of a point on the original figure, the corresponding point on the scaled figure has coordinates

$(x', y') = (mx, my).$

If m is greater than 1, the figure becomes larger; if m is between 0 and 1, it becomes smaller.

Shearing

A shearing transformation will push the top of a square sideways to form a parallelogram. Horizontal shearing is defined by:

$(x', y') = (x + ys, y)$

Shearing can also be applied vertically:

$(x', y') = (x, xs + y)$

9.5   Orientation and handedness

Main article: Orientation (mathematics)
See also: right-hand rule and Axes conventions

9.5.1   In two dimensions

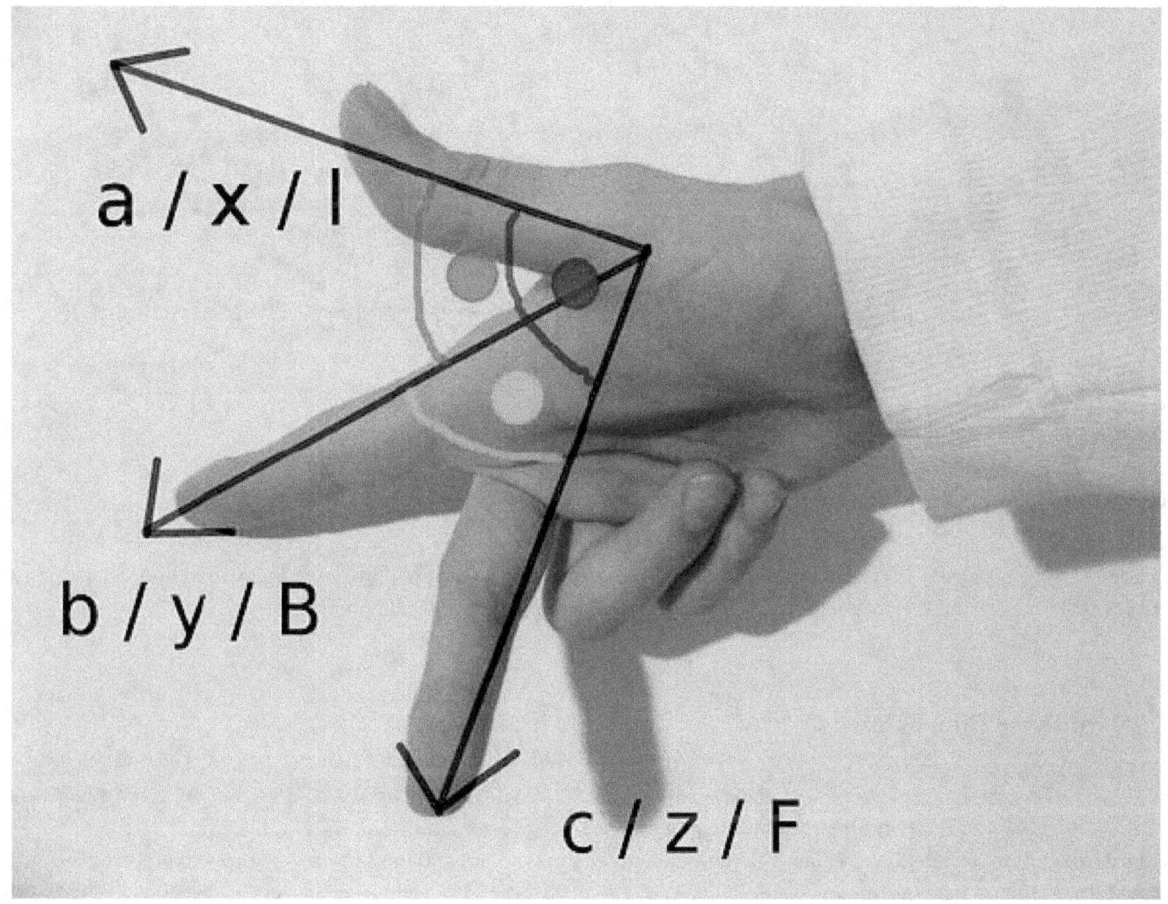

The right hand rule.

Fixing or choosing the x-axis determines the y-axis up to direction. Namely, the y-axis is necessarily the perpendicular to the x-axis through the point marked 0 on the x-axis. But there is a choice of which of the two half lines on the perpendicular to designate as positive and which as negative. Each of these two choices determines a different orientation (also called *handedness*) of the Cartesian plane.

The usual way of orienting the axes, with the positive x-axis pointing right and the positive y-axis pointing up (and the x-axis being the "first" and the y-axis the "second" axis) is considered the *positive* or *standard* orientation, also called the *right-handed* orientation.

A commonly used mnemonic for defining the positive orientation is the *right hand rule*. Placing a somewhat closed right hand on the plane with the thumb pointing up, the fingers point from the x-axis to the y-axis, in a positively oriented coordinate system.

The other way of orienting the axes is following the *left hand rule*, placing the left hand on the plane with the thumb pointing up.

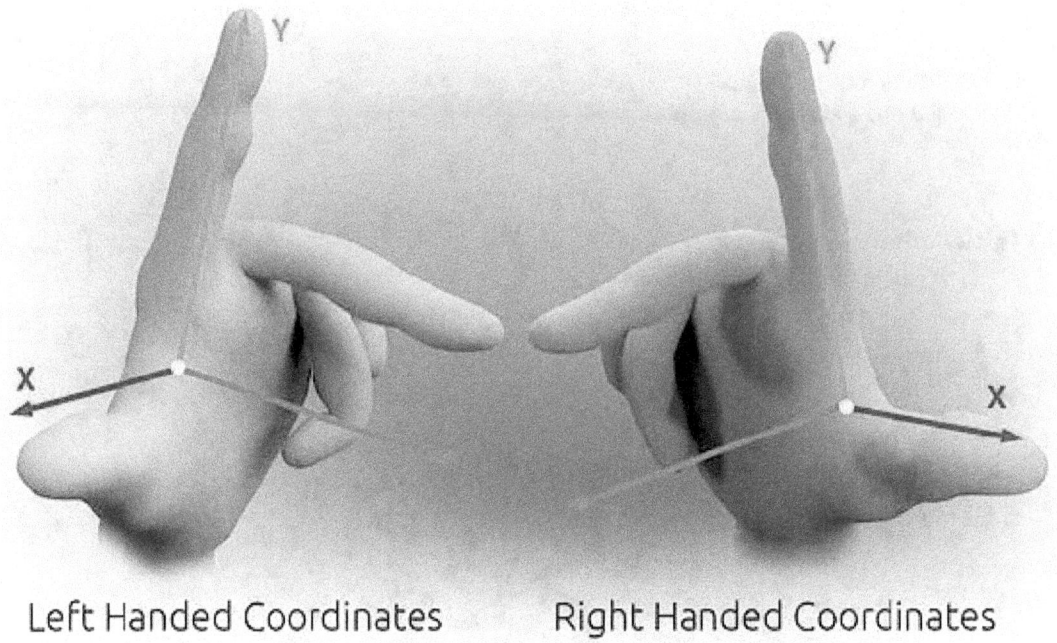

Left Handed Coordinates   Right Handed Coordinates

3D Cartesian Coordinate Handedness

When pointing the thumb away from the origin along an axis towards positive, the curvature of the fingers indicates a positive rotation along that axis.

Regardless of the rule used to orient the axes, rotating the coordinate system will preserve the orientation. Switching any two axes will reverse the orientation, but switching both will leave the orientation unchanged.

9.5.2   In three dimensions

Once the x- and y-axes are specified, they determine the line along which the z-axis should lie, but there are two possible directions on this line. The two possible coordinate systems which result are called 'right-handed' and 'left-handed'. The standard orientation, where the xy-plane is horizontal and the z-axis points up (and the x- and the y-axis form a positively oriented two-dimensional coordinate system in the xy-plane if observed from *above* the xy-plane) is called **right-handed** or **positive**.

The name derives from the right-hand rule. If the index finger of the right hand is pointed forward, the middle finger bent inward at a right angle to it, and the thumb placed at a right angle to both, the three fingers indicate the relative directions of the x-, y-, and z-axes in a *right-handed* system. The thumb indicates the x-axis, the index finger the y-axis and the middle finger the z-axis. Conversely, if the same is done with the left hand, a left-handed system results.

Figure 7 depicts a left and a right-handed coordinate system. Because a three-dimensional object is represented on the two-dimensional screen, distortion and ambiguity result. The axis pointing downward (and to the right) is also meant to point *towards* the observer, whereas the "middle" axis is meant to point *away* from the observer. The red circle is *parallel* to the horizontal xy-plane and indicates rotation from the x-axis to the y-axis (in both cases). Hence the red arrow passes *in front of* the z-axis.

Figure 8 is another attempt at depicting a right-handed coordinate system. Again, there is an ambiguity caused by projecting the three-dimensional coordinate system into the plane. Many observers see Figure 8 as "flipping in and out" between a convex cube and a concave "corner". This corresponds to the two possible orientations of the coordinate system. Seeing the figure as convex gives a left-handed coordinate system. Thus the "correct" way to view Figure 8 is to imagine the x-axis as pointing *towards* the observer and thus seeing a concave corner.

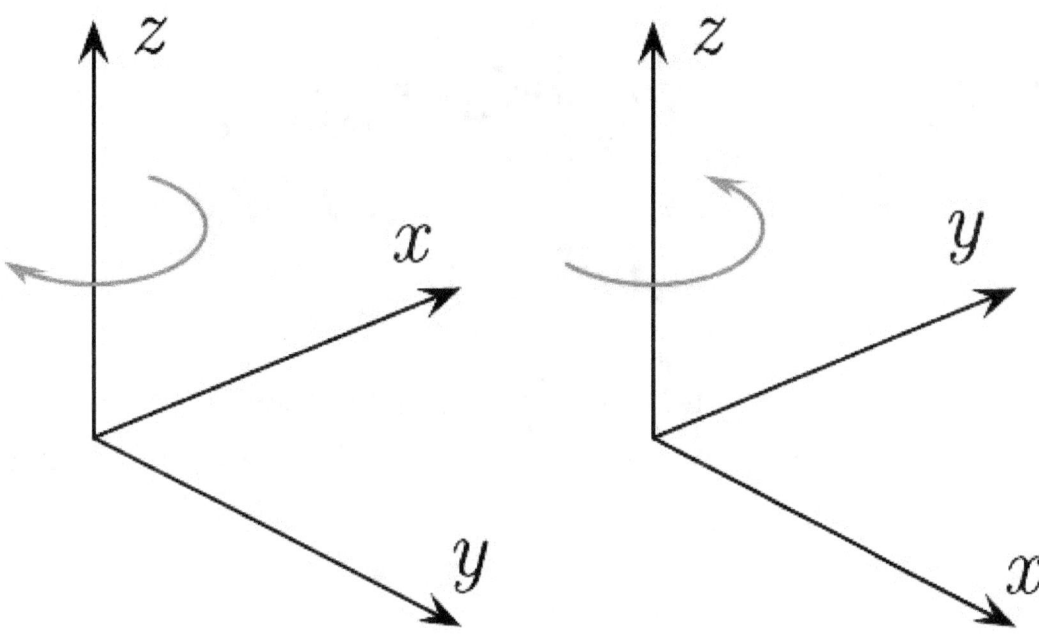

Fig. 7 – The left-handed orientation is shown on the left, and the right-handed on the right.

9.6   Representing a vector in the standard basis

A point in space in a Cartesian coordinate system may also be represented by a position vector, which can be thought of as an arrow pointing from the origin of the coordinate system to the point.[7] If the coordinates represent spatial positions (displacements), it is common to represent the vector from the origin to the point of interest as **r** . In two dimensions, the vector from the origin to the point with Cartesian coordinates (x, y) can be written as:

$$\mathbf{r} = x\mathbf{i} + y\mathbf{j}$$

where $\mathbf{i} = \begin{pmatrix} 1 \\ 0 \end{pmatrix}$, and $\mathbf{j} = \begin{pmatrix} 0 \\ 1 \end{pmatrix}$ are unit vectors in the direction of the x-axis and y-axis respectively, generally referred to as the *standard basis* (in some application areas these may also be referred to as versors). Similarly, in three dimensions, the vector from the origin to the point with Cartesian coordinates (x, y, z) can be written as:[8]

$$\mathbf{r} = x\mathbf{i} + y\mathbf{j} + z\mathbf{k}$$

where $\mathbf{k} = \begin{pmatrix} 0 \\ 0 \\ 1 \end{pmatrix}$ is the unit vector in the direction of the z-axis.

There is no *natural* interpretation of multiplying vectors to obtain another vector that works in all dimensions, however there is a way to use complex numbers to provide such a multiplication. In a two dimensional cartesian plane, identify the point with coordinates (x, y) with the complex number $z = x + iy$. Here, **i** is the imaginary unit and is identified with the point with coordinates (0, 1), so it is **not** the unit vector in the direction of the x-axis. Since the complex numbers can be multiplied giving another complex number, this identification provides a means to "multiply" vectors. In a three dimensional cartesian space a similar identification can be made with a subset of the quaternions.

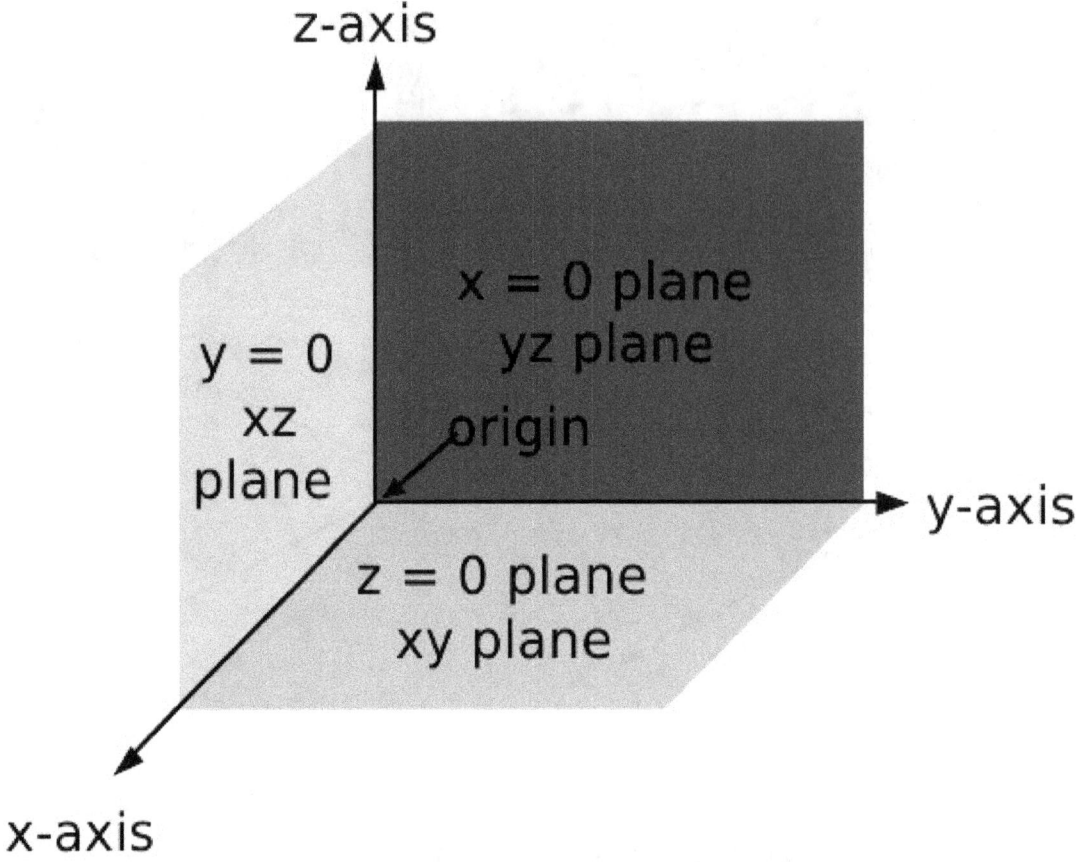

Fig. 8 – The right-handed Cartesian coordinate system indicating the coordinate planes.

9.7 Applications

Cartesian coordinates are an abstraction that have a multitude of possible applications in the real world. However, three constructive steps are involved in superimposing coordinates on a problem application. 1) Units of distance must be decided defining the spatial size represented by the numbers used as coordinates. 2) An origin must be assigned to a specific spatial location or landmark, and 3) the orientation of the axes must be defined using available directional cues for (n-1) of the n axes.

Consider as an example superimposing 3D Cartesian coordinates over all points on the Earth (i.e. geospatial 3D). What units make sense? Kilometers are a good choice, since the original definition of the kilometer was geospatial...10,000 km equalling the surface distance from the Equator to the North Pole. Where to place the origin? Based on symmetry, the gravitational center of the Earth suggests a natural landmark (which can be sensed via satellite orbits). Finally, how to orient X, Y and Z axis directions? The axis of Earth's spin provides a natural direction strongly associated with "up vs. down", so positive Z can adopt the direction from geocenter to North Pole. A location on the Equator is needed to define the X-axis, and the Prime Meridian stands out as a reference direction, so the X-axis takes the direction from geocenter out to [0 degrees longitude, 0 degrees latitude]. Note that with 3 dimensions, and two perpendicular axes directions pinned down for X and Z, the Y-axis is determined by the first two choices. In order to obey the right hand rule, the Y-axis must point out from the geocenter to [90 degrees longitude, 0 degrees latitude]. So what are the geocentric coordinates of the Empire State Building in New York City? Using [longitude = −73.985656, latitude = 40.748433], Earth radius = 40,000/2π, and transforming from spherical --> Cartesian coordinates, you can estimate the geocentric coordinates of the Empire State Building, [x, y, z] = [1330.53 km, −4635.75 km, 4155.46 km]. GPS navigation relies on such geocentric coordinates.

In engineering projects, agreement on the definition of coordinates is a crucial foundation. One cannot assume that coordinates come predefined for a novel application, so knowledge of how to erect a coordinate system where there is none is essential to applying René Descartes' ingenious thinking.

While spatial apps employ identical units along all axes, in business and scientific apps, each axis may have different units of measurement associated with it (such as kilograms, seconds, pounds, etc.). Although four- and higher-dimensional spaces are difficult to visualize, the algebra of Cartesian coordinates can be extended relatively easily to four or more variables, so that certain calculations involving many variables can be done. (This sort of algebraic extension is what is used to define the geometry of higher-dimensional spaces.) Conversely, it is often helpful to use the geometry of Cartesian coordinates in two or three dimensions to visualize algebraic relationships between two or three of many non-spatial variables.

The graph of a function or relation is the set of all points satisfying that function or relation. For a function of one variable, f, the set of all points (x, y), where $y = f(x)$ is the graph of the function f. For a function g of two variables, the set of all points (x, y, z), where $z = g(x, y)$ is the graph of the function g. A sketch of the graph of such a function or relation would consist of all the salient parts of the function or relation which would include its relative extrema, its concavity and points of inflection, any points of discontinuity and its end behavior. All of these terms are more fully defined in calculus. Such graphs are useful in calculus to understand the nature and behavior of a function or relation.

9.8 See also

- Horizontal and vertical

- Jones diagram, which plots four variables rather than two.

- Orthogonal coordinates

- Polar coordinate system

- Spherical coordinate system

9.9 Notes

9.10 References

[1] "Analytic geometry". *Encyclopædia Britannica* (Encyclopædia Britannica Online ed.). 2008.

[2] Burton 2011, p. 374

[3] A Tour of the Calculus, David Berlinski

[4] Springer online reference Encyclopedia of Mathematics

[5] Smart 1998, Chap. 2

[6] Brannan, Esplen & Gray 1998, pg. 49

[7] Brannan, Esplen & Gray 1998, Appendix 2, pp. 377–382

[8] David J. Griffiths (1999). *Introduction to Electrodynamics*. Prentice Hall. ISBN 0-13-805326-X.

9.11 Sources

- Brannan, David A.; Esplen, Matthew F.; Gray, Jeremy J. (1998), *Geometry*, Cambridge: Cambridge University Press, ISBN 0-521-59787-0

- Burton, David M. (2011), *The History of Mathematics/An Introduction* (7th ed.), New York: McGraw-Hill, ISBN 978-0-07-338315-6

- Smart, James R. (1998), *Modern Geometries* (5th ed.), Pacific Grove: Brooks/Cole, ISBN 0-534-35188-3

9.12 Further reading

- Descartes, René (2001). *Discourse on Method, Optics, Geometry, and Meteorology.* Trans. by Paul J. Oscamp (Revised ed.). Indianapolis, IN: Hackett Publishing. ISBN 0-87220-567-3. OCLC 488633510.

- Korn GA, Korn TM (1961). *Mathematical Handbook for Scientists and Engineers* (1st ed.). New York: McGraw-Hill. pp. 55–79. LCCN 59-14456. OCLC 19959906.

- Margenau H, Murphy GM (1956). *The Mathematics of Physics and Chemistry.* New York: D. van Nostrand. LCCN 55-10911.

- Moon P, Spencer DE (1988). "Rectangular Coordinates (x, y, z)". *Field Theory Handbook, Including Coordinate Systems, Differential Equations, and Their Solutions* (corrected 2nd, 3rd print ed.). New York: Springer-Verlag. pp. 9–11 (Table 1.01). ISBN 978-0-387-18430-2.

- Morse PM, Feshbach H (1953). *Methods of Theoretical Physics, Part I.* New York: McGraw-Hill. ISBN 0-07-043316-X. LCCN 52-11515.

- Sauer R, Szabó I (1967). *Mathematische Hilfsmittel des Ingenieurs.* New York: Springer Verlag. LCCN 67-25285.

9.13 External links

- Cartesian Coordinate System

- Printable Cartesian Coordinates

- Cartesian coordinates at PlanetMath.org.

- MathWorld description of Cartesian coordinates

- Coordinate Converter – converts between polar, Cartesian and spherical coordinates

- Coordinates of a point Interactive tool to explore coordinates of a point

- open source JavaScript class for 2D/3D Cartesian coordinate system manipulation

Chapter 10

Hyperbolic coordinates

In mathematics, **hyperbolic coordinates** are a method of locating points in quadrant I of the Cartesian plane

$$\{(x, y) : x > 0, \; y > 0\} = Q$$

Hyperbolic coordinates take values in the hyperbolic plane defined as:

$$HP = \{(u, v) : u \in \mathbb{R}, v > 0\}$$

These coordinates in HP are useful for studying logarithmic comparisons of direct proportion in Q and measuring deviations from direct proportion.

For (x, y) in Q take

$$u = \ln \sqrt{\frac{x}{y}}$$

and

$$v = \sqrt{xy}$$

Sometimes the parameter u is called the hyperbolic angle and v Is called the geometric mean.

The inverse mapping is

$$x = ve^u, \quad y = ve^{-u}$$

This is a continuous mapping, but not an analytic function.

10.1 Alternative quadrant metric

Since HP carries the metric space structure of the Poincaré half-plane model of hyperbolic geometry, the bijective correspondence $Q \leftrightarrow HP$ brings this structure to Q. It can be grasped using the notion of hyperbolic motions. Since geodesics in HP are semicircles with centers on the boundary, the geodesics in Q are obtained from the correspondence and turn

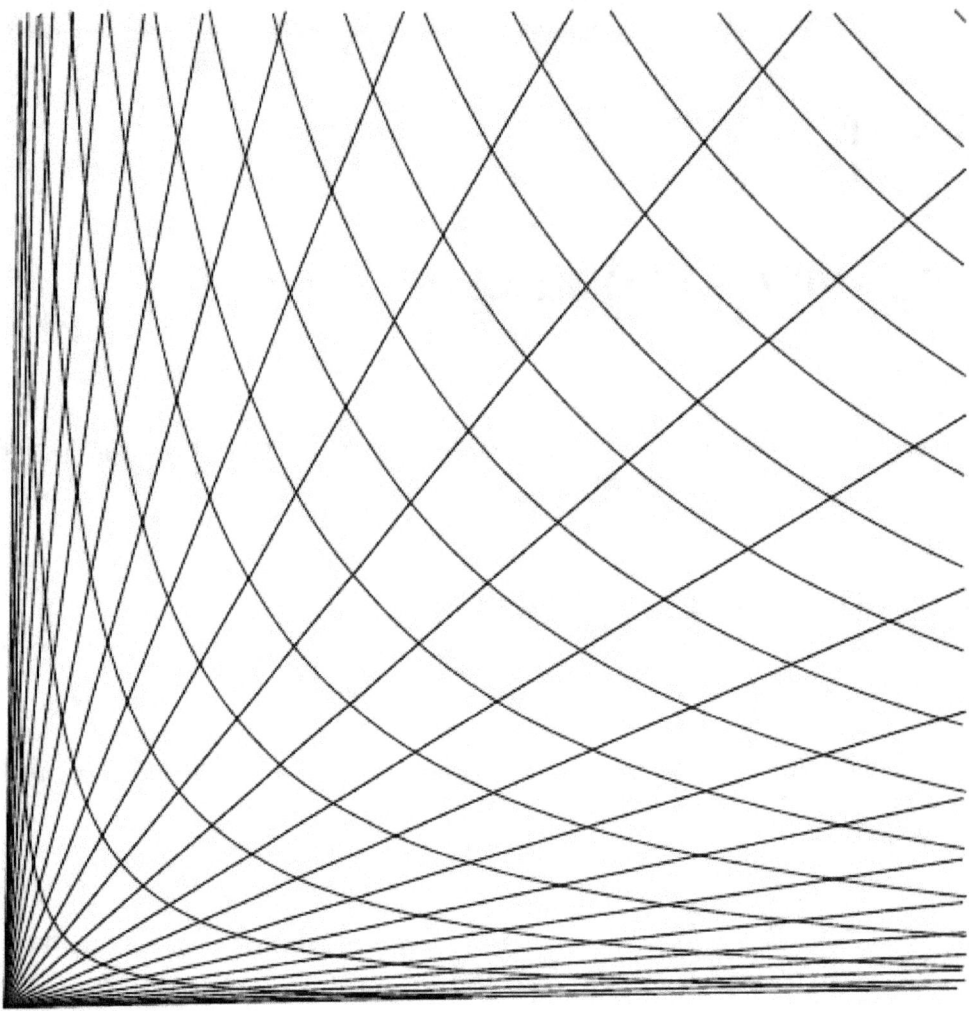

Hyperbolic coordinates plotted on the Euclidean plane: all points on the same blue ray share the same coordinate value u, *and all points on the same red hyperbola share the same coordinate value* v.

out to be rays from the origin or petal-shaped curves leaving and re-entering the origin. And the hyperbolic motion of *HP* given by a left-right shift corresponds to a squeeze mapping applied to *Q*.

Since hyperbolas in *Q* correspond to lines parallel to the boundary of *HP*, they are horocycles in the metric geometry of *Q*.

If one only considers the Euclidean topology of the plane and the topology inherited by *Q*, then the lines bounding *Q* seem close to *Q*. Insight from the metric space *HP* shows that the open set *Q* has only the origin as boundary when viewed through the correspondence. Indeed, consider rays from the origin in *Q*, and their images, vertical rays from the boundary *R* of *HP*. Any point in *HP* is an infinite distance from the point *p* at the foot of the perpendicular to *R*, but a sequence of points on this perpendicular may tend in the direction of *p*. The corresponding sequence in *Q* tends along a ray toward the origin. The old Euclidean boundary of *Q* is no longer relevant.

10.2   Applications in physical science

Fundamental physical variables are sometimes related by equations of the form $k = x\,y$. For instance, $V = I\,R$ (Ohm's law), $P = V\,I$ (electrical power), $P\,V = k\,T$ (ideal gas law), and $f\,\lambda = v$ (relation of wavelength, frequency, and velocity in the wave medium). When the k is constant, the other variables lie on a hyperbola, which is a horocycle in the appropriate Q quadrant.

For example, in thermodynamics the isothermal process explicitly follows the hyperbolic path and work can be interpreted as a hyperbolic angle change. Similarly, a given mass M of gas with changing volume will have variable density $\delta = M/V$, and the ideal gas law may be written $P = k\,T\,\delta$ so that an isobaric process traces a hyperbola in the quadrant of absolute temperature and gas density.

For hyperbolic coordinates in the theory of relativity see the History section.

10.3   Statistical applications

- Comparative study of population density in the quadrant begins with selecting a reference nation, region, or urban area whose population and area are taken as the point $(1,1)$.

- Analysis of the elected representation of regions in a representative democracy begins with selection of a standard for comparison: a particular represented group, whose magnitude and slate magnitude (of representatives) stands at $(1,1)$ in the quadrant.

10.4   Economic applications

There are many natural applications of hyperbolic coordinates in economics:

- Analysis of currency exchange rate fluctuation:

The unit currency sets $x = 1$. The price currency corresponds to y. For

$$0 < y < 1$$

we find $u > 0$, a positive hyperbolic angle. For a *fluctuation* take a new price

$$0 < z < y$$

Then the change in u is:

$$\Delta u = \ln\sqrt{\frac{y}{z}}$$

Quantifying exchange rate fluctuation through hyperbolic angle provides an objective, symmetric, and consistent measure. The quantity Δu is the length of the left-right shift in the hyperbolic motion view of the currency fluctuation.

- Analysis of inflation or deflation of prices of a basket of consumer goods.

- Quantification of change in marketshare in duopoly.

- Corporate stock splits versus stock buy-back.

10.5   History

While the geometric mean is an ancient concept, the hyperbolic angle is contemporary with the development of logarithm, the latter part of the seventeenth century. Gregoire de Saint-Vincent, Marin Mersenne, and Alphonse Antonio de Sarasa evaluated the quadrature of the hyperbola as a function having properties now familiar for the logarithm. The exponential function, the hyperbolic sine, and the hyperbolic cosine followed. As complex function theory referred to infinite series the circular functions sine and cosine seemed to absorb the hyperbolic sine and cosine as depending on an imaginary variable. In the nineteenth century biquaternions came into use and exposed the alternative complex plane called split-complex numbers where the hyperbolic angle is raised to a level equal to the classical angle. In English literature biquaternions were used to model spacetime and show its symmetries. There the hyperbolic angle parameter came to be called rapidity. For relativists, examining the quadrant as the possible future between oppositely directed photons, the geometric mean parameter is temporal.

In relativity the focus is on the 3-dimensional hypersurface in the future of spacetime where various velocities arrive after a given proper time. Scott Walter[1] explains that in November 1907 Hermann Minkowski alluded to a well-known three-dimensional hyperbolic geometry while speaking to the Göttingen Mathematical Society, but not to a four-dimensional one.[2] In tribute to Wolfgang Rindler, the author of the standard introductory university-level textbook on relativity, hyperbolic coordinates of spacetime are called Rindler coordinates.

10.6   References

[1]   Walter (1999) page 6

[2]   Walter (1999) page 8

- David Betounes (2001) *Differential Equations: Theory and Applications*, page 254, Springer-TELOS, ISBN 0-387-95140-7 .

- Scott Walter (1999). "The non-Euclidean style of Minkowskian relativity". Chapter 4 in: Jeremy J. Gray (ed.), *The Symbolic Universe: Geometry and Physics 1890-1930*, pp. 91–127. Oxford University Press. ISBN 0-19-850088-2.

Chapter 11

Hyperbola

This article is about a geometric curve. For the term used in rhetoric, see Hyperbole.

In mathematics, a **hyperbola** (plural *hyperbolas* or *hyperbolae*) is a type of smooth curve, lying in a plane, defined by its geometric properties or by equations for which it is the solution set. A hyperbola has two pieces, called connected components or branches, that are mirror images of each other and resemble two infinite bows. The hyperbola is one of the four kinds of conic section, formed by the intersection of a plane and a double cone. (The other conic sections are the parabola, the ellipse, and the circle; the circle is a special case of the ellipse). If the plane intersects both halves of the double cone but does not pass through the apex of the cones, then the conic is a hyperbola.

Hyperbolas arise in many ways: as the curve representing the function $f(x) = 1/x$ in the Cartesian plane, as the appearance of a circle viewed from within it, as the path followed by the shadow of the tip of a sundial, as the shape of an open orbit (as distinct from a closed elliptical orbit), such as the orbit of a spacecraft during a gravity assisted swing-by of a planet or more generally any spacecraft exceeding the escape velocity of the nearest planet, as the path of a single-apparition comet (one travelling too fast ever to return to the solar system), as the scattering trajectory of a subatomic particle (acted on by repulsive instead of attractive forces but the principle is the same), and so on.

Each branch of the hyperbola has two arms which become straighter (lower curvature) further out from the center of the hyperbola. Diagonally opposite arms, one from each branch, tend in the limit to a common line, called the asymptote of those two arms. So there are two asymptotes, whose intersection is at the center of symmetry of the hyperbola, which can be thought of as the mirror point about which each branch reflects to form the other branch. In the case of the curve $f(x) = 1/x$ the asymptotes are the two coordinate axes.

Hyperbolas share many of the ellipses' analytical properties such as eccentricity, focus, and directrix. Typically the correspondence can be made with nothing more than a change of sign in some term. Many other mathematical objects have their origin in the hyperbola, such as hyperbolic paraboloids (saddle surfaces), hyperboloids ("wastebaskets"), hyperbolic geometry (Lobachevsky's celebrated non-Euclidean geometry), hyperbolic functions (sinh, cosh, tanh, etc.), and gyrovector spaces (a geometry used in both relativity and quantum mechanics which is not Euclidean).

11.1   History

The word "hyperbola" derives from the Greek ὑπερβολή, meaning "over-thrown" or "excessive", from which the English term hyperbole also derives. Hyperbolae were discovered by Menaechmus in his investigations of the problem of doubling the cube, but were then called sections of obtuse cones.[1] The term hyperbola is believed to have been coined by Apollonius of Perga (c. 262–c. 190 BC) in his definitive work on the conic sections, the *Conics*.[2] For comparison, the other two general conic sections, the ellipse and the parabola, derive from the corresponding Greek words for "deficient" and "comparable"; these terms may refer to the eccentricity of these curves, which is greater than one (hyperbola), less than one (ellipse) and exactly one (parabola).

11.2 Nomenclature and features

Similar to a parabola, a hyperbola is an open curve, meaning that it continues indefinitely to infinity, rather than closing on itself as an ellipse does. A hyperbola consists of two disconnected curves called its **arms** or **branches**.

The points on the two branches that are closest to each other are called the vertices; they are the points where the curve has its smallest radius of curvature. The line segment connecting the vertices is called the *transverse axis* or *major axis*, corresponding to the major diameter of an ellipse. The midpoint of the transverse axis is known as the hyperbola's *center*. The distance a from the center to each vertex is called the semi-major axis. Outside of the transverse axis but on the same line are the two *focal points (foci)* of the hyperbola. The line through these five points is one of the two principal axes of the hyperbola, the other being the perpendicular bisector of the transverse axis. The hyperbola has mirror symmetry about its principal axes, and is also symmetric under a $180°$ turn about its center.

At large distances from the center, the hyperbola approaches two lines, its asymptotes, which intersect at the hyperbola's center. A hyperbola approaches its asymptotes arbitrarily closely as the distance from its center increases, but it never intersects them; however, a degenerate hyperbola consists only of its asymptotes. Consistent with the symmetry of the hyperbola, if the transverse axis is aligned with the x-axis of a Cartesian coordinate system, the slopes of the asymptotes are equal in magnitude but opposite in sign, $\pm^b/_a$, where $b=a\times\tan(\theta)$ and where θ is the angle between the transverse axis and either asymptote. The distance b (not shown) is the length of the perpendicular segment from either vertex to the asymptotes.

A *conjugate axis* of length $2b$, corresponding to the *minor axis* of an ellipse, is sometimes drawn on the non-transverse principal axis; its endpoints $\pm b$ lie on the minor axis at the height of the asymptotes over/under the hyperbola's vertices. Because of the minus sign in some of the formulas below, it is also called the *imaginary axis* of the hyperbola.

If $b = a$, the angle 2θ between the asymptotes equals $90°$ and the hyperbola is said to be *rectangular* or *equilateral*. In this special case, the rectangle joining the four points on the asymptotes directly above and below the vertices is a square, since the lengths of its sides $2a = 2b$.

If the transverse axis of any hyperbola is aligned with the x-axis of a Cartesian coordinate system and is centered on the origin, the equation of the hyperbola can be written as

$$\frac{x^2}{a^2} - \frac{y^2}{b^2} = 1.$$

A hyperbola aligned in this way is called an "East-West opening hyperbola". Likewise, a hyperbola with its transverse axis aligned with the y-axis is called a "North–South opening hyperbola" and has equation

$$\frac{y^2}{a^2} - \frac{x^2}{b^2} = 1.$$

Every hyperbola is congruent to the origin-centered East-West opening hyperbola sharing its same eccentricity ε (its shape, or degree of "spread"), and is also congruent to the origin-centered North–South opening hyperbola with identical eccentricity ε — that is, it can be rotated so that it opens in the desired direction and can be translated (rigidly moved in the plane) so that it is centered at the origin. For convenience, hyperbolas are usually analyzed in terms of their centered East-West opening form.

If c is the distance from the center to either focus, then $a^2 + b^2 = c^2$.

The shape of a hyperbola is defined entirely by its eccentricity ε, which is a dimensionless number always greater than one. The distance c from the center to the foci equals $a\varepsilon$. The eccentricity can also be defined as the ratio of the distances to either focus and to a corresponding line known as the directrix; hence, the distance from the center to the directrices equals a/ε. In terms of the parameters a, b, c and the angle θ, the eccentricity equals

$$\varepsilon = \frac{c}{a} = \frac{\sqrt{a^2 + b^2}}{a} = \sqrt{1 + \frac{b^2}{a^2}} = \sec\theta.$$

For example, the eccentricity of a rectangular hyperbola ($\theta = 45°$, $a = b$) equals the square root of two: $\varepsilon = \sqrt{2}$.

Every hyperbola has a *conjugate hyperbola*, in which the transverse and conjugate axes are exchanged without changing the asymptotes. The equation of the conjugate hyperbola of $\frac{x^2}{a^2} - \frac{y^2}{b^2} = 1$ is $\frac{x^2}{a^2} - \frac{y^2}{b^2} = -1$. If the graph of the conjugate hyperbola is rotated $90°$ to restore the east-west opening orientation (so that x becomes y and vice versa), the equation of the resulting rotated conjugate hyperbola is the same as the equation of the original hyperbola except with a and b exchanged. For example, the angle θ of the conjugate hyperbola equals $90°$ minus the angle of the original hyperbola. Thus, the angles in the original and conjugate hyperbolas are complementary angles, which implies that they have different eccentricities unless $\theta = 45°$ (a rectangular hyperbola). Hence, the conjugate hyperbola does *not* in general correspond to a $90°$ rotation of the original hyperbola; the two hyperbolas are generally different in shape.

A few other lengths are used to describe hyperbolas. Consider a line perpendicular to the transverse axis (i.e., parallel to the conjugate axis) that passes through one of the hyperbola's foci. The line segment connecting the two intersection points of this line with the hyperbola is known as the *latus rectum* and has a length $\frac{2b^2}{a}$. The *semi-latus rectum* l is half of this length, i.e., $l = \frac{b^2}{a}$. The *focal parameter* p is the distance from a focus to its corresponding directrix, and equals $p = \frac{b^2}{c}$.

11.3   Mathematical definitions

A hyperbola can be defined mathematically in several equivalent ways.

11.3.1   Conic section

A hyperbola may be defined as the curve of intersection between a right circular conical surface and a plane that cuts through both halves of the cone. The other major types of conic sections are the ellipse and the parabola; in these cases, the plane cuts through only one half of the double cone. If the plane passes through the central apex of the double cone a degenerate hyperbola results — two straight lines that cross at the apex point.

11.3.2   Difference of distances to foci

A hyperbola may be defined equivalently as the locus of points where the absolute value of the *difference* of the distances to the two foci is a constant equal to $2a$, the distance between its two vertices. This definition accounts for many of the hyperbola's applications, such as multilateration; this is the problem of determining position from the *difference* in arrival times of synchronized signals, as in GPS.

This definition may be expressed also in terms of tangent circles. The center of any circles externally tangent to two given circles lies on a hyperbola, whose foci are the centers of the given circles and where the vertex distance $2a$ equals the difference in radii of the two circles. As a special case, one given circle may be a point located at one focus; since a point may be considered as a circle of zero radius, the other given circle—which is centered on the other focus—must have radius $2a$. This provides a simple technique for constructing a hyperbola, as shown below. It follows from this definition that a tangent line to the hyperbola at a point $\mathbf{P}$ bisects the angle formed with the two foci, i.e., the angle $\mathbf{F}_1\mathbf{P}\mathbf{F}_2$. Consequently, the feet of perpendiculars drawn from each focus to such a tangent line lies on a circle of radius a that is centered on the hyperbola's own center.

A proof that this characterization of the hyperbola is equivalent to the conic-section characterization can be done without coordinate geometry by means of Dandelin spheres.

11.3.3   Directrix and focus

A hyperbola can be defined as the locus of points for which the ratio of the distances to one focus and to a line (called the directrix) is a constant ϵ that is larger than 1. This constant is the eccentricity of the hyperbola. The eccentricity equals the secant of half the angle between the asymptotes of the hyperbola, so the eccentricity of the hyperbola xy = 1 equals the square root of 2.

By symmetry a hyperbola has two directrices, which are parallel to the conjugate axis and are between it and the tangent to the hyperbola at a vertex. One directrix and its focus is enough to produce both arms of the hyperbola.

11.3.4 Reciprocation of a circle

The reciprocation of a circle B in a circle C always yields a conic section such as a hyperbola. The process of "reciprocation in a circle C" consists of replacing every line and point in a geometrical figure with their corresponding pole and polar, respectively. The *pole* of a line is the inversion of its closest point to the circle C, whereas the polar of a point is the converse, namely, a line whose closest point to C is the inversion of the point.

The eccentricity of the conic section obtained by reciprocation is the ratio of the distances between the two circles' centers to the radius r of reciprocation circle C. If $\mathbf{B}$ and $\mathbf{C}$ represent the points at the centers of the corresponding circles, then

$$\epsilon = \frac{\overline{BC}}{r}$$

Since the eccentricity of a hyperbola is always greater than one, the center $\mathbf{B}$ must lie outside of the reciprocating circle C.

This definition implies that the hyperbola is both the locus of the poles of the tangent lines to the circle B, as well as the envelope of the polar lines of the points on B. Conversely, the circle B is the envelope of polars of points on the hyperbola, and the locus of poles of tangent lines to the hyperbola. Two tangent lines to B have no (finite) poles because they pass through the center $\mathbf{C}$ of the reciprocation circle C; the polars of the corresponding tangent points on B are the asymptotes of the hyperbola. The two branches of the hyperbola correspond to the two parts of the circle B that are separated by these tangent points.

11.3.5 Quadratic equation

A hyperbola can also be defined as a second-degree equation in the Cartesian coordinates (x, y) of the plane

$$A_{xx}x^2 + 2A_{xy}xy + A_{yy}y^2 + 2B_x x + 2B_y y + C = 0$$

provided that the constants A_{xx}, A_{xy}, A_{yy}, B_x, B_y, and C satisfy the determinant condition

$$D = \begin{vmatrix} A_{xx} & A_{xy} \\ A_{xy} & A_{yy} \end{vmatrix} < 0$$

A special case of a hyperbola—the *degenerate hyperbola* consisting of two intersecting lines—occurs when another determinant is zero

$$\Delta := \begin{vmatrix} A_{xx} & A_{xy} & B_x \\ A_{xy} & A_{yy} & B_y \\ B_x & B_y & C \end{vmatrix} = 0$$

This determinant Δ is sometimes called the discriminant of the conic section.[3]

Given the above general parametrization of the hyperbola in Cartesian coordinates, the eccentricity can be found using the formula in Conic section#Eccentricity in terms of parameters of the quadratic form.

The center (x_c, y_c) of the hyperbola may be determined from the formulae

$$x_c = -\frac{1}{D}\begin{vmatrix} B_x & A_{xy} \\ B_y & A_{yy} \end{vmatrix}$$

$$y_c = -\frac{1}{D}\begin{vmatrix} A_{xx} & B_x \\ A_{xy} & B_y \end{vmatrix}$$

In terms of new coordinates, $\xi = x - xc$ and $\eta = y - yc$, the defining equation of the hyperbola can be written

$$A_{xx}\xi^2 + 2A_{xy}\xi\eta + A_{yy}\eta^2 + \frac{\Delta}{D} = 0$$

The principal axes of the hyperbola make an angle Φ with the positive x-axis that equals

$$\tan 2\Phi = \frac{2A_{xy}}{A_{xx} - A_{yy}}$$

Rotating the coordinate axes so that the x-axis is aligned with the transverse axis brings the equation into its **canonical form**

$$\frac{x^2}{a^2} - \frac{y^2}{b^2} = 1$$

The major and minor semiaxes a and b are defined by the equations

$$a^2 = -\frac{\Delta}{\lambda_1 D} = -\frac{\Delta}{\lambda_1^2 \lambda_2}$$

$$b^2 = -\frac{\Delta}{\lambda_2 D} = -\frac{\Delta}{\lambda_1 \lambda_2^2}$$

where λ_1 and λ_2 are the roots of the quadratic equation

$$\lambda^2 - (A_{xx} + A_{yy})\lambda + D = 0$$

For comparison, the corresponding equation for a degenerate hyperbola is

$$\frac{x^2}{a^2} - \frac{y^2}{b^2} = 0$$

The tangent line to a given point (x_0, y_0) on the hyperbola is defined by the equation

$$Ex + Fy + G = 0$$

where E, F and G are defined

$$E = A_{xx}x_0 + A_{xy}y_0 + B_x$$

$$F = A_{xy}x_0 + A_{yy}y_0 + B_y$$

$$G = B_x x_0 + B_y y_0 + C$$

The normal line to the hyperbola at the same point is given by the equation

$$F(x - x_0) - E(y - y_0) = 0$$

The normal line is perpendicular to the tangent line, and both pass through the same point (x_0, y_0).

From the equation

$$\frac{x^2}{a^2} - \frac{y^2}{b^2} = 1 \qquad 0 < b \le a$$

the basic property that with r_1 and r_2 being the distances from a point (x, y) to the left focus $(-ae, 0)$ and the right focus $(ae, 0)$ one has for a point on the right branch that

$$r_1 - r_2 = 2a$$

and for a point on the left branch that

$$r_2 - r_1 = 2a$$

can be proved as follows:

If x,y is a point on the hyperbola the distance to the left focal point is

$$r_1^2 = (x + ae)^2 + y^2 = x^2 + 2xae + a^2 e^2 + (x^2 - a^2)(e^2 - 1) = (ex + a)^2$$

To the right focal point the distance is

$$r_2^2 = (x - ae)^2 + y^2 = x^2 - 2xae + a^2 e^2 + (x^2 - a^2)(e^2 - 1) = (ex - a)^2$$

If x,y is a point on the right branch of the hyperbola then $ex > a$ and

$$r_1 = ex + a$$

$$r_2 = ex - a$$

Subtracting these equations one gets

$$r_1 - r_2 = 2a$$

If x,y is a point on the left branch of the hyperbola then $ex < -a$ and

$$r_1 = -ex - a$$

$$r_2 = -ex + a$$

Subtracting these equations one gets

$$r_2 - r_1 = 2a$$

11.4   True anomaly

In the section above it is shown that using the coordinate system in which the equation of the hyperbola takes its **canonical form**

$$\frac{x^2}{a^2} - \frac{y^2}{b^2} = 1$$

the distance r from a point (x , y) on the left branch of the hyperbola to the left focal point $(-ea , 0)$ is

$$r = -ex - a$$

Introducing polar coordinates (r , θ) with origin at the left focal point the coordinates relative the canonical coordinate system are

$$x = -ae + r\cos\theta$$

$$y = r\sin\theta$$

and the equation above takes the form

$$r = -e(-ae + r\cos\theta) - a$$

from which follows that

$$r = \frac{a(e^2 - 1)}{1 + e\cos\theta}$$

This is the representation of the near branch of a hyperbola in polar coordinates with respect to a focal point.

The polar angle θ of a point on a hyperbola relative the near focal point as described above is called the **true anomaly** of the point.

11.5   Geometrical constructions

Similar to the ellipse, a hyperbola can be constructed using a taut thread. A straightedge of length S is attached to one focus F_1 at one of its corners A so that it is free to rotate about that focus. A thread of length $L = S - 2a$ is attached between the other focus F_2 and the other corner B of the straightedge. A sharp pencil is held up against the straightedge, sandwiching the thread tautly against the straightedge. Let the position of the pencil be denoted as P. The total length L of the thread equals the sum of the distances L_2 from F_2 to P and LB from P to B. Similarly, the total length S of the straightedge equals the distance L_1 from F_1 to P and LB. Therefore, the difference in the distances to the foci, $L_1 - L_2$ equals the constant $2a$

$$L_1 - L_2 = (S - L_B) - (L - L_B) = S - L = 2a$$

A second construction uses intersecting circles, but is likewise based on the constant difference of distances to the foci. Consider a hyperbola with two foci F_1 and F_2, and two vertices P and Q; these four points all lie on the transverse axis. Choose a new point T also on the transverse axis and to the right of the rightmost vertex P; the difference in distances to the two vertices, QT − PT = 2*a*, since 2*a* is the distance between the vertices. Hence, the two circles centered on the foci F_1 and F_2 of radius QT and PT, respectively, will intersect at two points of the hyperbola.

A third construction relies on the definition of the hyperbola as the reciprocation of a circle. Consider the circle centered on the center of the hyperbola and of radius *a*; this circle is tangent to the hyperbola at its vertices. A line *g* drawn from one focus may intersect this circle in two points **M** and **N**; perpendiculars to *g* drawn through these two points are tangent to the hyperbola. Drawing a set of such tangent lines reveals the envelope of the hyperbola.

A fourth construction is using the parallelogram method. It is similar to such method for parabola and ellipse construction: certain equally spaced points lying on parallel lines are connected with each other by two straight lines and their intersection point lies on the hyperbola.

11.6 Reflections and tangent lines

The ancient Greek geometers recognized a reflection property of hyperbolas. If a ray of light emerges from one focus and is reflected from either branch of the hyperbola, the light-ray appears to have come from the other focus. Equivalently, by reversing the direction of the light, rays directed at one of the foci are reflected towards the other focus. This property is analogous to the property of ellipses that a ray emerging from one focus is reflected from the ellipse directly *towards* the other focus (rather than *away* as in the hyperbola). Expressed mathematically, lines drawn from each focus to the same point on the hyperbola intersect it at equal angles; the tangent line to a hyperbola at a point **P** bisects the angle formed with the two foci, F_1PF_2.

Tangent lines to a hyperbola have another remarkable geometrical property. If a tangent line at a point **T** intersects the asymptotes at two points **K** and **L**, then **T** bisects the line segment KL, and the product of distances to the hyperbola's center, OK×OL is a constant.

11.7 Hyperbolic functions and equations

Just as the sine and cosine functions give a parametric equation for the ellipse, so the hyperbolic sine and hyperbolic cosine give a parametric equation for the hyperbola.

As

$$\cosh^2 \mu - \sinh^2 \mu = 1$$

one has for any hyperbolic angle μ that the point

$$x = a \cosh \mu$$

$$y = b \sinh \mu$$

satisfies the equation

$$\frac{x^2}{a^2} - \frac{y^2}{b^2} = 1$$

which is the equation of a hyperbola relative its canonical coordinate system.

When μ varies over the interval $-\infty < \mu < \infty$ one gets with this formula all points (x, y) on the right branch of the hyperbola.

The left branch for which $x < 0$ is in the same way obtained as

$$x = -a \cosh \mu$$

$y = b \sinh \mu$

In the figure the points (x_k , y_k) given by

$x_k = -a \cosh \mu_k$

$y_k = b \sinh \mu_k$

for

$\mu_k = 0.3 k \quad k = -5, -4, \cdots, 5$

on the left branch of a hyperbola with eccentricity 1.2 are marked as dots.

11.8   Relation to other conic sections

There are three major types of conic sections: hyperbolas, ellipses and parabolas. Since the parabola may be seen as a limiting case poised exactly between an ellipse and a hyperbola, there are effectively only two major types, ellipses and hyperbolas. These two types are related in that formulae for one type can often be applied to the other.

The canonical equation for a hyperbola is

$$\frac{x^2}{a^2} - \frac{y^2}{b^2} = 1.$$

Any hyperbola can be rotated so that it is east-west opening and positioned with its center at the origin, so that the equation describing it is this canonical equation.

The canonical equation for the hyperbola may be seen as a version of the corresponding ellipse equation

$$\frac{x^2}{a^2} + \frac{y^2}{b^2} = 1$$

in which the semi-minor axis length b is imaginary. That is, if in the ellipse equation b is replaced by ib where b is real, one obtains the hyperbola equation.

Similarly, the parametric equations for a hyperbola and an ellipse are expressed in terms of hyperbolic and trigonometric functions, respectively, which are again related by an imaginary circular angle, for example,

$\cosh \mu = \cos i\mu$

Hence, many formulae for the ellipse can be extended to hyperbolas by adding the imaginary unit i in front of the semi-minor axis b and the angle. For example, the arc length of a segment of an ellipse can be determined using an incomplete elliptic integral of the second kind. The corresponding arclength of a hyperbola is given by the same function with imaginary parameters b and μ, namely, $ib\,E(i\mu, c)$.

11.9   Conic section analysis of the hyperbolic appearance of circles

Besides providing a uniform description of circles, ellipses, parabolas, and hyperbolas, conic sections can also be understood as a natural model of the geometry of perspective in the case where the scene being viewed consists of a circle, or more generally an ellipse. The viewer is typically a camera or the human eye. In the simplest case the viewer's lens is just a pinhole; the role of more complex lenses is merely to gather far more light while retaining as far as possible the simple pinhole geometry in which all rays of light from the scene pass through a single point. Once through the lens, the rays then spread out again, in air in the case of a camera, in the vitreous humor in the case of the eye, eventually distributing themselves over the film, imaging device, or retina, all of which come under the heading of image plane. The **lens plane** is a plane parallel to the image plane at the lens; all rays pass through a single point on the lens plane, namely the lens itself.

When the circle directly faces the viewer, the viewer's lens is on-axis, meaning on the line normal to the circle through its center (think of the axle of a wheel). The rays of light from the circle through the lens to the image plane then form a cone with circular cross section whose apex is the lens. The image plane concretely realizes the abstract cutting plane in the conic section model.

When in addition the viewer directly faces the circle, the circle is rendered faithfully on the image plane without perspective distortion, namely as a scaled-down circle. When the viewer turns attention or gaze away from the center of the circle the image plane then cuts the cone in an ellipse, parabola, or hyperbola depending on how far the viewer turns, corresponding exactly to what happens when the surface cutting the cone to form a conic section is rotated.

A parabola arises when the lens plane is tangent to (touches) the circle. A viewer with perfect 180-degree wide-angle vision will see the whole parabola; in practice this is impossible and only a finite portion of the parabola is captured on the film or retina.

When the viewer turns further so that the lens plane cuts the circle in two points, the shape on the image plane becomes that of a hyperbola. The viewer still sees only a finite curve, namely a portion of one branch of the hyperbola, and is unable to see the second branch at all, which corresponds to the portion of the circle behind the viewer, more precisely, on the same side of the lens plane as the viewer. In practice the finite extent of the image plane makes it impossible to see any portion of the circle near where it is cut by the lens plane. Further back however one could imagine rays from the portion of the circle well behind the viewer passing through the lens, were the viewer transparent. In this case the rays would pass through the image plane before the lens, yet another impracticality ensuring that no portion of the second branch could possibly be visible.

The tangents to the circle where it is cut by the lens plane constitute the asymptotes of the hyperbola. Were these tangents to be drawn in ink in the plane of the circle, the eye would perceive them as asymptotes to the visible branch. Whether they converge in front of or behind the viewer depends on whether the lens plane is in front of or behind the center of the circle respectively.

If the circle is drawn on the ground and the viewer gradually transfers gaze from straight down at the circle up towards the horizon, the lens plane eventually cuts the circle producing first a parabola then a hyperbola on the image plane as shown in Figure 10. As the gaze continues to rise the asymptotes of the hyperbola, if realized concretely, appear coming in from left and right, swinging towards each other and converging at the horizon when the gaze is horizontal. Further elevation of the gaze into the sky then brings the point of convergence of the asymptotes towards the viewer.

By the same principle with which the back of the circle appears on the image plane were all the physical obstacles to its projection to be overcome, the portion of the two tangents behind the viewer appear on the image plane as an extension of the visible portion of the tangents in front of the viewer. Like the second branch this extension materializes in the sky rather than on the ground, with the horizon marking the boundary between the physically visible (scene in front) and invisible (scene behind), and the visible and invisible parts of the tangents combining in a single X shape. As the gaze is raised and lowered about the horizon, the X shape moves oppositely, lowering as the gaze is raised and vice versa but always with the visible portion being on the ground and stopping at the horizon, with the center of the X being on the horizon when the gaze is horizontal.

All of the above was for the case when the circle faces the viewer, with only the viewer's gaze varying. When the circle starts to face away from the viewer the viewer's lens is no longer on-axis. In this case the cross section of the cone is no longer a circle but an ellipse (never a parabola or hyperbola). However the principle of conic sections does not depend on the cross section of the cone being circular, and applies without modification to the case of eccentric cones.

It is not difficult to see that even in the off-axis case a circle can appear circular, namely when the image plane (and hence lens plane) is parallel to the plane of the circle. That is, to see a circle as a circle when viewing it obliquely, look not at the circle itself but at the plane in which it lies. From this it can be seen that when viewing a plane filled with many circles, all of them will appear circular simultaneously when the plane is looked at directly.

A common misperception about the hyperbola is that it is a mathematical curve rarely if ever encountered in daily life. The reality is that one sees a hyperbola whenever catching sight of portion of a circle cut by one's lens plane (and a parabola when the lens plane is tangent to, i.e. just touches, the circle). The inability to see very much of the arms of the visible branch, combined with the complete absence of the second branch, makes it virtually impossible for the human visual system to recognize the connection with hyperbolas such as $y = 1/x$ where both branches are on display simultaneously.

11.10   Derived curves

Several other curves can be derived from the hyperbola by inversion, the so-called inverse curves of the hyperbola. If the center of inversion is chosen as the hyperbola's own center, the inverse curve is the lemniscate of Bernoulli; the lemniscate is also the envelope of circles centered on a rectangular hyperbola and passing through the origin. If the center of inversion is chosen at a focus or a vertex of the hyperbola, the resulting inverse curves are a limaçon or a strophoid, respectively.

11.11   Coordinate systems

11.11.1   Cartesian coordinates

An east-west opening hyperbola centered at (h,k) has the equation

$$\frac{(x-h)^2}{a^2} - \frac{(y-k)^2}{b^2} = 1.$$

The major axis runs through the center of the hyperbola and intersects both arms of the hyperbola at the vertices (bend points) of the arms. The foci lie on the extension of the major axis of the hyperbola.

The minor axis runs through the center of the hyperbola and is perpendicular to the major axis.

In both formulas a is the semi-major axis (half the distance between the two arms of the hyperbola measured along the major axis),[4] and b is the semi-minor axis (half the distance between the asymptotes along a line tangent to the hyperbola at a vertex).

If one forms a rectangle with vertices on the asymptotes and two sides that are tangent to the hyperbola, the sides tangent to the hyperbola are $2b$ in length while the sides that run parallel to the line between the foci (the major axis) are $2a$ in length. Note that b may be larger than a despite the names *minor* and *major*.

If one calculates the distance from any point on the hyperbola to each focus, the absolute value of the difference of those two distances is always $2a$.

The eccentricity is given by

$$\varepsilon = \sqrt{1 + \frac{b^2}{a^2}} = \sec\left(\arctan\left(\frac{b}{a}\right)\right) = \cosh\left(\operatorname{arcsinh}\left(\frac{b}{a}\right)\right)$$

If c equals the distance from the center to either focus, then

$$\varepsilon = \frac{c}{a}$$

where

$$c = \sqrt{a^2 + b^2}$$

The distance c is known as the **linear eccentricity** of the hyperbola. The distance between the foci is $2c$ or $2ae$.

The foci for an east-west opening hyperbola are given by

$$(h \pm c, k)$$

and for a north-south opening hyperbola are given by

$$(h, k \pm c)$$

The directrices for an east-west opening hyperbola are given by

$$x = h \pm a \, \cos\left(\arctan\left(\frac{b}{a}\right)\right)$$

and for a north-south opening hyperbola are given by

$$y = k \pm a \, \cos\left(\arctan\left(\frac{b}{a}\right)\right)$$

11.11.2   Polar coordinates

The polar coordinates used most commonly for the hyperbola are defined relative to the Cartesian coordinate system that has its origin in a focus and its x-axis pointing towards the origin of the "canonical coordinate system" as illustrated in the figure of the section "True anomaly".

Relative to this coordinate system one has that

$$r = \frac{a(e^2 - 1)}{1 + e \cos \theta}$$

and the range of the true anomaly θ is:

$$-\arccos\left(-\frac{1}{e}\right) < \theta < \arccos\left(-\frac{1}{e}\right)$$

With polar coordinate relative to the "canonical coordinate system"

$$x = R \cos t$$
$$y = R \sin t$$

one has that

$$R^2 = \frac{b^2}{e^2 \cos^2 t - 1}$$

For the right branch of the hyperbola the range of t is:

$$-\arccos\left(\frac{1}{e}\right) < t < \arccos\left(\frac{1}{e}\right)$$

11.11.3   Parametric equations

East-west opening hyperbola:

$$x = a \sec t + h \qquad x = \pm a \cosh t + h$$
$$y = b \tan t + k \quad \text{or} \quad y = b \sinh t + k$$

North-south opening hyperbola:

$$x = b \tan t + h \qquad x = b \sinh t + h$$
$$y = a \sec t + k \quad \text{or} \quad y = \pm a \cosh t + k$$

In all formulae (h,k) are the center coordinates of the hyperbola, a is the length of the semi-major axis, and b is the length of the semi-minor axis.

11.11.4   Elliptic coordinates

A family of confocal hyperbolas is the basis of the system of elliptic coordinates in two dimensions. These hyperbolas are described by the equation

$$\left(\frac{x}{c \cos \theta}\right)^2 - \left(\frac{y}{c \sin \theta}\right)^2 = 1$$

where the foci are located at a distance c from the origin on the x-axis, and where θ is the angle of the asymptotes with the x-axis. Every hyperbola in this family is orthogonal to every ellipse that shares the same foci. This orthogonality may be shown by a conformal map of the Cartesian coordinate system $w = z + 1/z$, where $z = x + iy$ are the original Cartesian coordinates, and $w = u + iv$ are those after the transformation.

Other orthogonal two-dimensional coordinate systems involving hyperbolas may be obtained by other conformal mappings. For example, the mapping $w = z^2$ transforms the Cartesian coordinate system into two families of orthogonal hyperbolas.

11.12   Rectangular hyperbola

A **rectangular hyperbola**, **equilateral hyperbola**, or **right hyperbola** is a hyperbola for which the asymptotes are perpendicular.[5]

Rectangular hyperbolas with the coordinate axes parallel to their asymptotes have the equation

$$(x - h)(y - k) = m$$

Rectangular hyperbolas have eccentricity $\varepsilon = \sqrt{2}$ with semi-major axis and semi-minor axis given by $a = b = \sqrt{2m}$.

The simplest example of rectangular hyperbolas occurs when the center (h, k) is at the origin:

$$y = \frac{m}{x}$$

describing quantities x and y that are inversely proportional. By rotating the coordinate axes counterclockwise by 45 degrees, with the new coordinate axes labelled (x', y') the equation of the hyperbola is given by canonical form

$$\frac{(x')^2}{(\sqrt{2m})^2} - \frac{(y')^2}{(\sqrt{2m})^2} = 1$$

If the scale factor $m=1/2$, then this canonical rectangular hyperbola is the unit hyperbola.

A circumconic passing through the orthocenter of a triangle is a rectangular hyperbola.[6]

11.13 Other properties of hyperbolas

- If a line intersects one branch of a hyperbola at M and N and intersects the asymptotes at P and Q, then MN has the same midpoint as PQ.[7][8]:p.49,ex.7

- The following are concurrent: (1) a circle passing through the hyperbola's foci and centered at the hyperbola's center; (2) either of the lines that are tangent to the hyperbola at the vertices; and (3) either of the asymptotes of the hyperbola.[7][9]

- The following are also concurrent: (1) the circle that is centered at the hyperbola's center and that passes through the hyperbola's vertices; (2) either directrix; and (3) either of the asymptotes.[9]

- The product of the distances from a point P on the hyperbola to one of the asymptotes along a line parallel to the other asymptote, and to the second asymptote along a line parallel to the first asymptote, is independent of the location of point P on the hyperbola.[9] If the hyperbola is written in canonical form $\frac{x^2}{a^2} - \frac{y^2}{b^2} = 1$ then this product is $\frac{a^2+b^2}{4}$.

- The product of the perpendicular distances from a point P on the hyperbola $\frac{x^2}{a^2} - \frac{y^2}{b^2} = 1$ or on its conjugate hyperbola $\frac{x^2}{a^2} - \frac{y^2}{b^2} = -1$ to the asymptotes is a constant independent of the location of P: specifically, $\frac{a^2 b^2}{a^2+b^2}$, which also equals $(b/e)^2$ where e is the eccentricity of the hyperbola $\frac{x^2}{a^2} - \frac{y^2}{b^2} = 1$.[10]

- The product of the slopes of lines from a point on the hyperbola to the two vertices is independent of the location of the point.[11]

- A line segment between the two asymptotes and tangent to the hyperbola is bisected by the tangency point.[8]:p.49,ex.6

- The area of a triangle two of whose sides lie on the asymptotes, and whose third side is tangent to the hyperbola, is independent of the location of the tangency point.[8]:p.49,ex.6 Specifically, the area is ab where a is the semi-major axis and b is the semi-minor axis.[12]

- The distance from either focus to either asymptote is b, the semi-minor axis; the nearest point to a focus on an asymptote lies at a distance from the center equal to a, the semi-major axis.[7] Then using the Pythagorean theorem on the right triangle with these two segments as legs shows that $a^2 + b^2 = c^2$, where c is the semi-focal length (the distance from a focus to the hyperbola's center).

11.14 Applications

11.14.1 Sundials

Hyperbolas may be seen in many sundials. On any given day, the sun revolves in a circle on the celestial sphere, and its rays striking the point on a sundial traces out a cone of light. The intersection of this cone with the horizontal plane of the ground forms a conic section. At most populated latitudes and at most times of the year, this conic section is a hyperbola. In practical terms, the shadow of the tip of a pole traces out a hyperbola on the ground over the course of a day (this path is called the *declination line*). The shape of this hyperbola varies with the geographical latitude and with the time of the year, since those factors affect the cone of the sun's rays relative to the horizon. The collection of such hyperbolas for a whole year at a given location was called a *pelekinon* by the Greeks, since it resembles a double-bladed axe.

11.14.2   Multilateration

A hyperbola is the basis for solving Multilateration problems, the task of locating a point from the differences in its distances to given points — or, equivalently, the difference in arrival times of synchronized signals between the point and the given points. Such problems are important in navigation, particularly on water; a ship can locate its position from the difference in arrival times of signals from a LORAN or GPS transmitters. Conversely, a homing beacon or any transmitter can be located by comparing the arrival times of its signals at two separate receiving stations; such techniques may be used to track objects and people. In particular, the set of possible positions of a point that has a distance difference of $2a$ from two given points is a hyperbola of vertex separation $2a$ whose foci are the two given points.

11.14.3   Path followed by a particle

The path followed by any particle in the classical Kepler problem is a conic section. In particular, if the total energy E of the particle is greater than zero (i.e., if the particle is unbound), the path of such a particle is a hyperbola. This property is useful in studying atomic and sub-atomic forces by scattering high-energy particles; for example, the Rutherford experiment demonstrated the existence of an atomic nucleus by examining the scattering of alpha particles from gold atoms. If the short-range nuclear interactions are ignored, the atomic nucleus and the alpha particle interact only by a repulsive Coulomb force, which satisfies the inverse square law requirement for a Kepler problem.

11.14.4   Korteweg-de Vries equation

The hyperbolic trig function sech x appears as one solution to the Korteweg-de Vries equation which describes the motion of a soliton wave in a canal.

11.14.5   Angle trisection

As shown first by Apollonius of Perga, a hyperbola can be used to trisect any angle, a well studied problem of geometry. Given an angle, first draw a circle centered at its vertex **O**, which intersects the sides of the angle at points **A** and **B**. Next draw the line through **A** and **B** and its perpendicular bisector ℓ. Construct a hyperbola of eccentricity $\varepsilon=2$ with ℓ as directrix and **B** as a focus. Let **P** be the intersection (upper) of the hyperbola with the circle. Angle **POB** trisects angle **AOB**. To prove this, reflect the line segment **OP** about the line ℓ obtaining the point **P'** as the image of **P**. Segment **AP'** has the same length as segment **BP** due to the reflection, while segment **PP'** has the same length as segment **BP** due to the eccentricity of the hyperbola. As **OA**, **OP'**, **OP** and **OB** are all radii of the same circle (and so, have the same length), the triangles **OAP'**, **OPP'** and **OPB** are all congruent. Therefore, the angle has been trisected, since 3×**POB** = **AOB**.[13]

11.14.6   Efficient portfolio frontier

In portfolio theory, the locus of mean-variance efficient portfolios (called the efficient frontier) is the upper half of the east-opening branch of a hyperbola drawn with the portfolio return's standard deviation plotted horizontally and its expected value plotted vertically; according to this theory, all rational investors would choose a portfolio characterized by some point on this locus.

11.15 Extensions

The three-dimensional analog of a hyperbola is a hyperboloid. Hyperboloids come in two varieties, those of one sheet and those of two sheets. A simple way of producing a hyperboloid is to rotate a hyperbola about the axis of its foci or about its symmetry axis perpendicular to the first axis: these rotations produce hyperboloids of two and one sheet, respectively.

11.16 See also

11.16.1 Other conic sections

- Circle
- Ellipse
- Parabola

11.16.2 Other related topics

- Apollonius of Perga, the Greek geometer who gave the ellipse, parabola, and hyperbola the names by which we know them.
- Elliptic coordinates, an orthogonal coordinate system based on families of ellipses and hyperbolas.
- Hyperbolic function
- Hyperbolic growth
- Hyperbolic partial differential equation
- Hyperbolic sector
- Hyperbolic structure
- Hyperbolic trajectory
- Hyperboloid
- Multilateration
- Rotation of axes
- Translation of axes
- Unit hyperbola

11.17 Notes

[1] Heath, Sir Thomas Little (1896), "Chapter I. The discovery of conic sections. Menaechmus", *Apollonius of Perga: Treatise on Conic Sections with Introductions Including an Essay on Earlier History on the Subject*, Cambridge University Press, pp. xvii–xxx.

[2] Boyer, Carl B.; Merzbach, Uta C. (2011), *A History of Mathematics*, Wiley, p. 73, ISBN 9780470630563, It was Apollonius (possibly following up a suggestion of Archimedes) who introduced the names "ellipse" and "hyperbola" in connection with these curves.

[3] Korn, Granino A. and Korn, Theresa M. *Mathematical Handbook for Scientists and Engineers: Definitions, Theorems, and Formulas for Reference and Review*, Dover Publ., second edition, 2000: p. 40.

[4] In some literature the value of *a* is taken negative for a hyperbola (the negative of half the distance between the two arms of the hyperbola measured along the major axis). This allows some formulas to be applicable to ellipses as well as to hyperbolas.

[5] Weisstein, Eric W. "Rectangular Hyperbola." From MathWorld--A Wolfram Web Resource. http://mathworld.wolfram.com/RectangularHyperbola.html

[6] Weisstein, Eric W. "Jerabek Hyperbola." From MathWorld--A Wolfram Web Resource. http://mathworld.wolfram.com/JerabekHyperbola.htm

[7]

[8] Spain, Barry. *Analytical Conics.* Dover Publ., 2007.

[9]

[10] Mitchell, Douglas W., "A property of hyperbolas and their asymptotes", *Mathematical Gazette* 96, July 2012, 299-301.

[11]

[12]

[13] This construction is due to Pappus of Alexandria (circa 300 A.D.) and the proof comes from Kazarinoff (1970, pg. 62).

11.18   References

- Kazarinoff, Nicholas D. (2003), *Ruler and the Round*, Mineola, N.Y.: Dover, ISBN 0-486-42515-0

11.19   External links

- Hazewinkel, Michiel, ed. (2001), "Hyperbola", *Encyclopedia of Mathematics*, Springer, ISBN 978-1-55608-010-4

- Apollonius' Derivation of the Hyperbola at Convergence

- Unit hyperbola at PlanetMath.org.

- Conic section at PlanetMath.org.

- Conjugate hyperbola at PlanetMath.org.

- Weisstein, Eric W., "Hyperbola", *MathWorld*.

A hyperbola is an open curve with two branches, the intersection of a plane with both halves of a double cone. The plane does not have to be parallel to the axis of the cone; the hyperbola will be symmetrical in any case.

Hyperbolas in the physical world: three cones of light of different widths and intensities are generated by a (roughly) downwards-pointing halogen lamp and its holder. Each cone of light intersects a nearby vertical wall in a hyperbola.

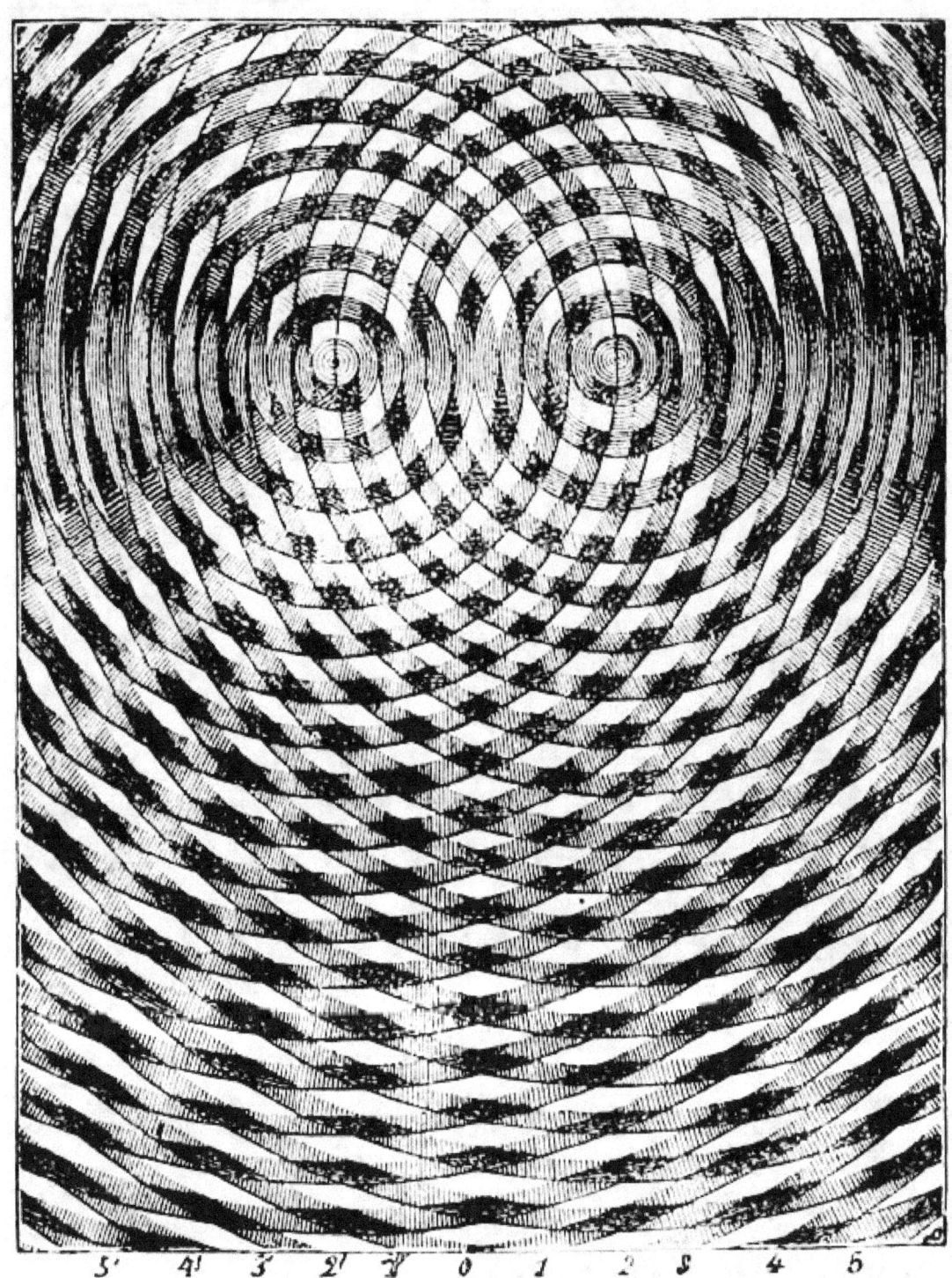

Hyperbolas produced by interference of waves

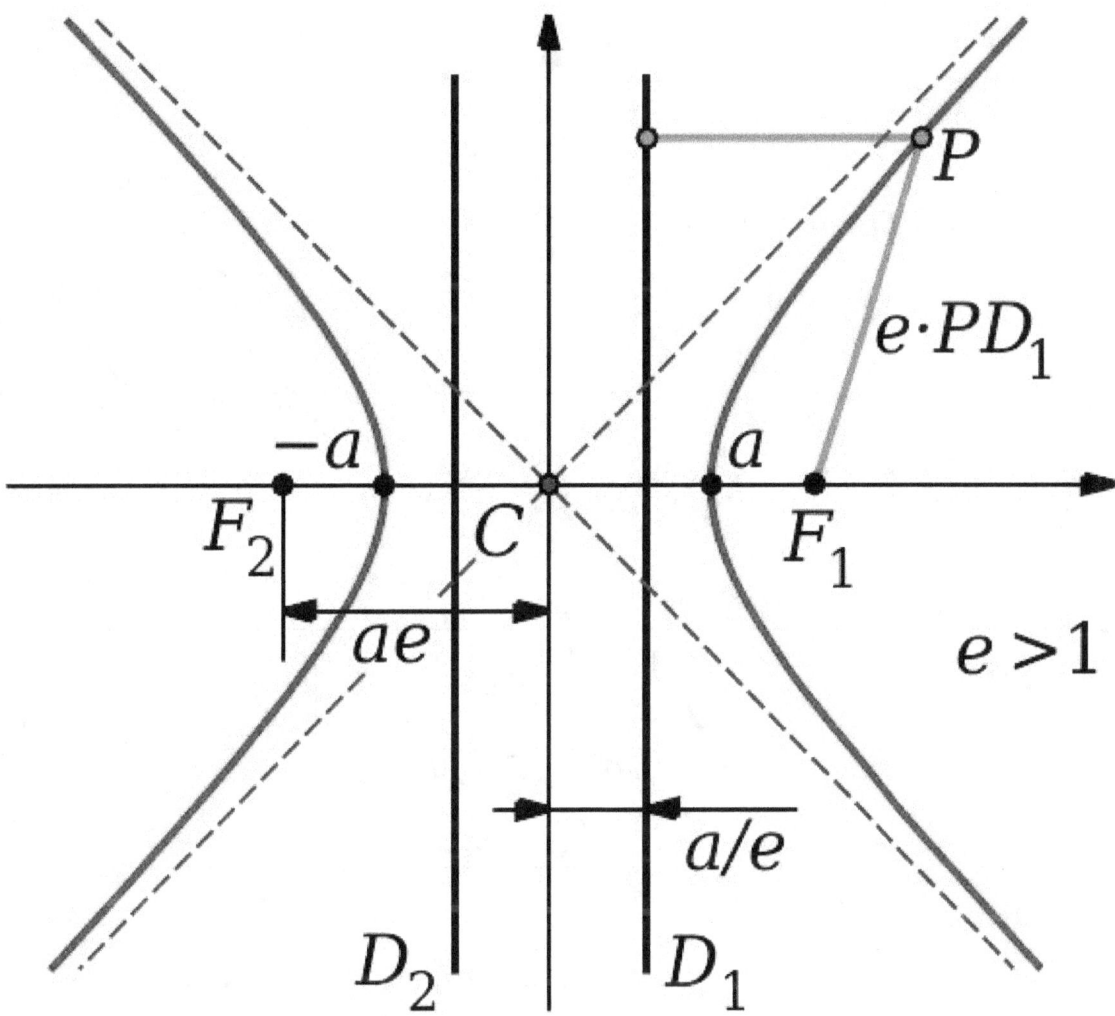

The hyperbola consists of the red curves. The asymptotes of the hyperbola are shown as blue dashed lines and intersect at the center of the hyperbola, C. The two focal points are labeled F_1 and F_2, and the thin black line joining them is the transverse axis. The perpendicular thin black line through the center is the conjugate axis. The two thick black lines parallel to the conjugate axis (thus, perpendicular to the transverse axis) are the two directrices, D_1 and D_2. The eccentricity e equals the ratio of the distances from a point P on the hyperbola to one focus and its corresponding directrix line (shown in green). The two vertices are located on the transverse axis at ±a relative to the center. So the parameters are: a — distance from center C to either vertex

b — length of a segment perpendicular to the transverse axis drawn from each vertex to the asymptotes

c — distance from center C to either Focus point, F_1 and F_2, and

θ — angle formed by each asymptote with the transverse axis.

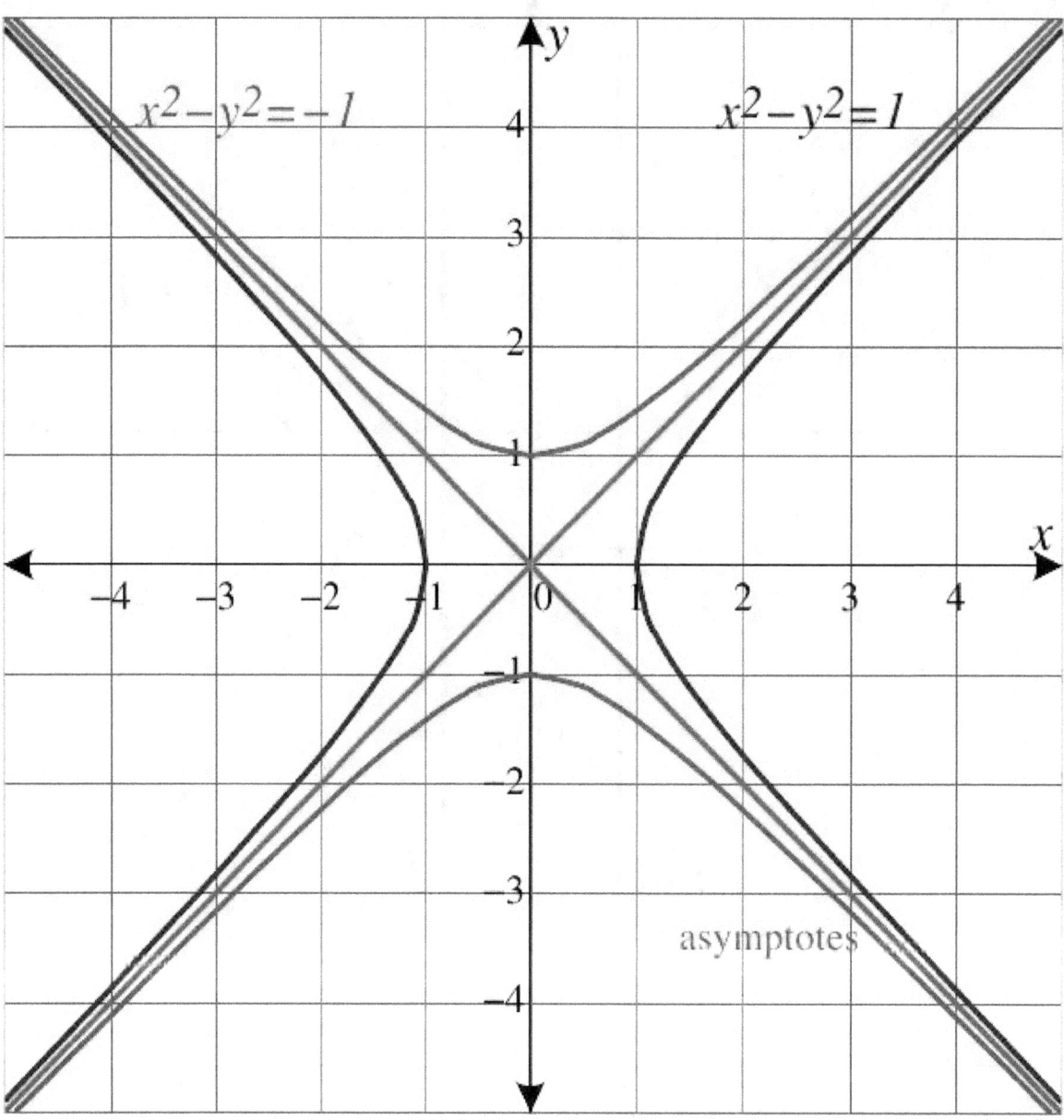

Here a = b = *1 giving the unit hyperbola in blue and its conjugate hyperbola in green, sharing the same red asymptotes.*

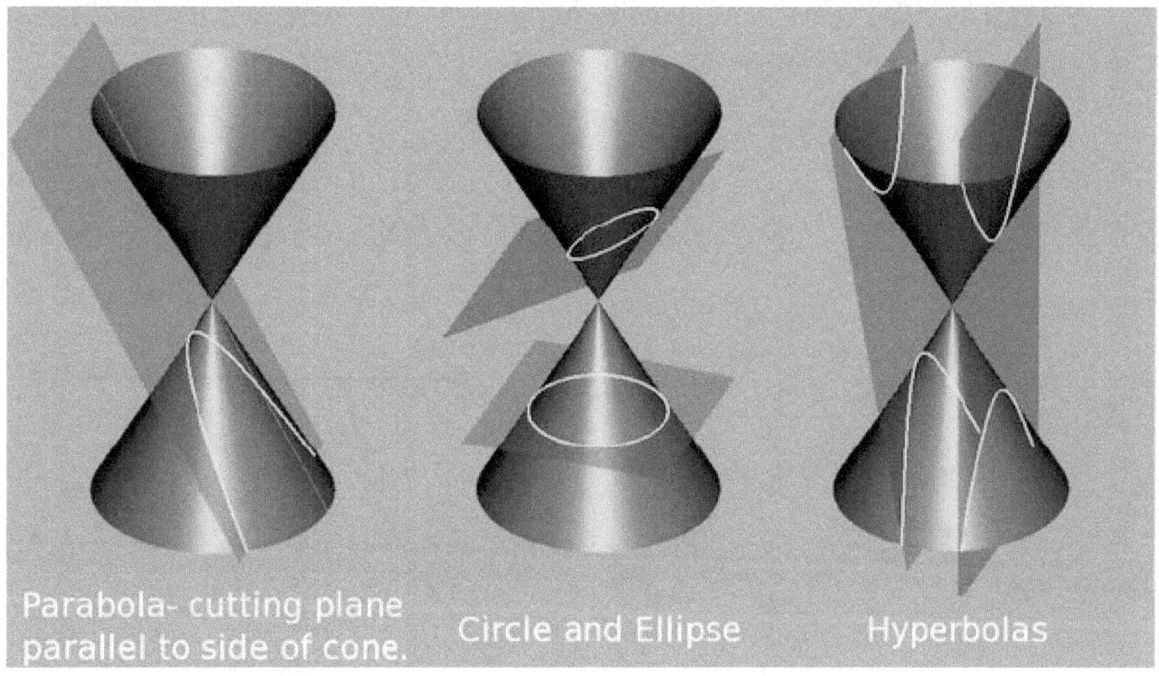

Three major types of conic sections.

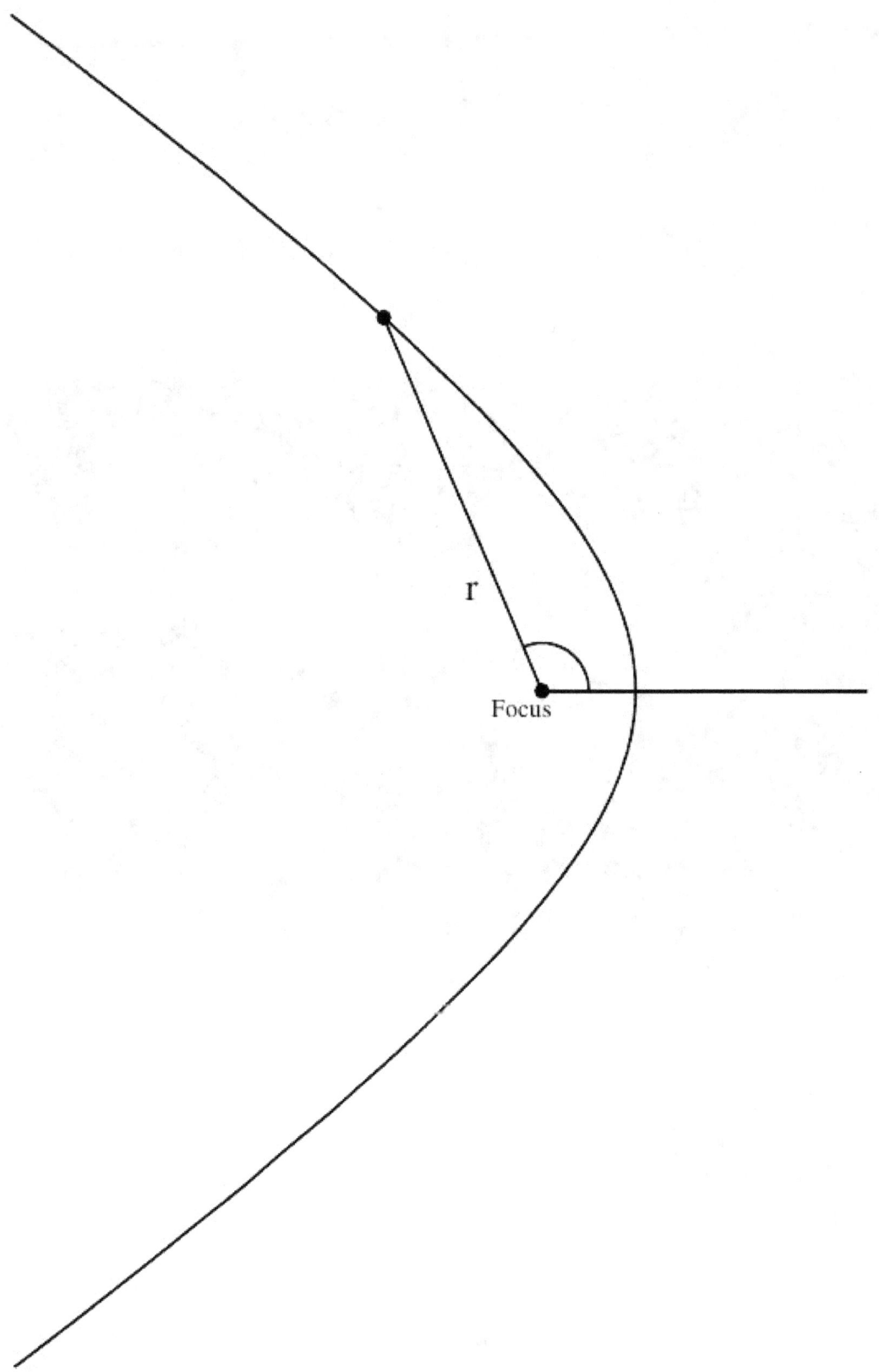

The angle shown is the true anomaly of the indicated point on the hyperbola.

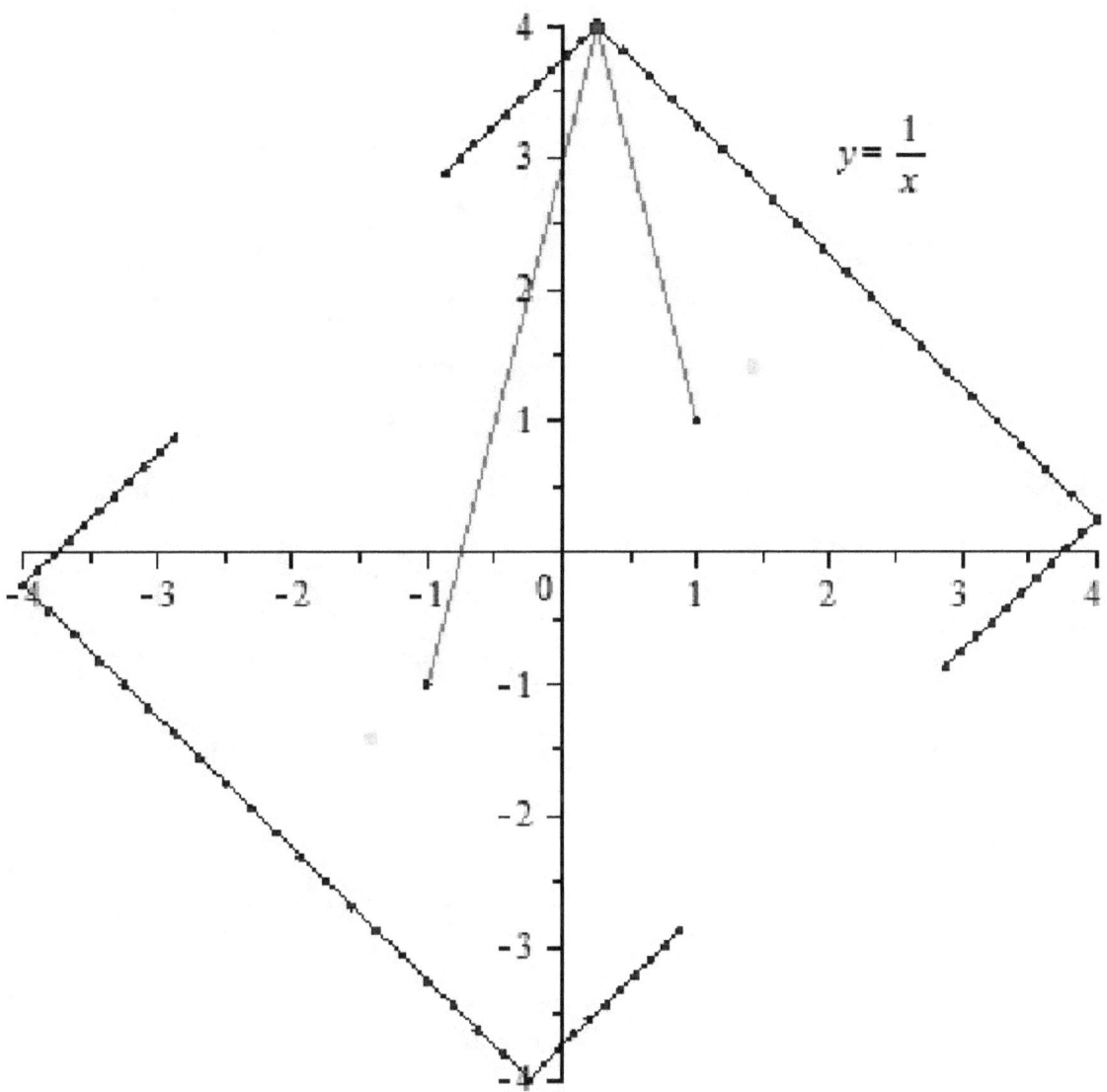

$$y = \frac{1}{x}$$

Hyperbola construction using the parallelogram method

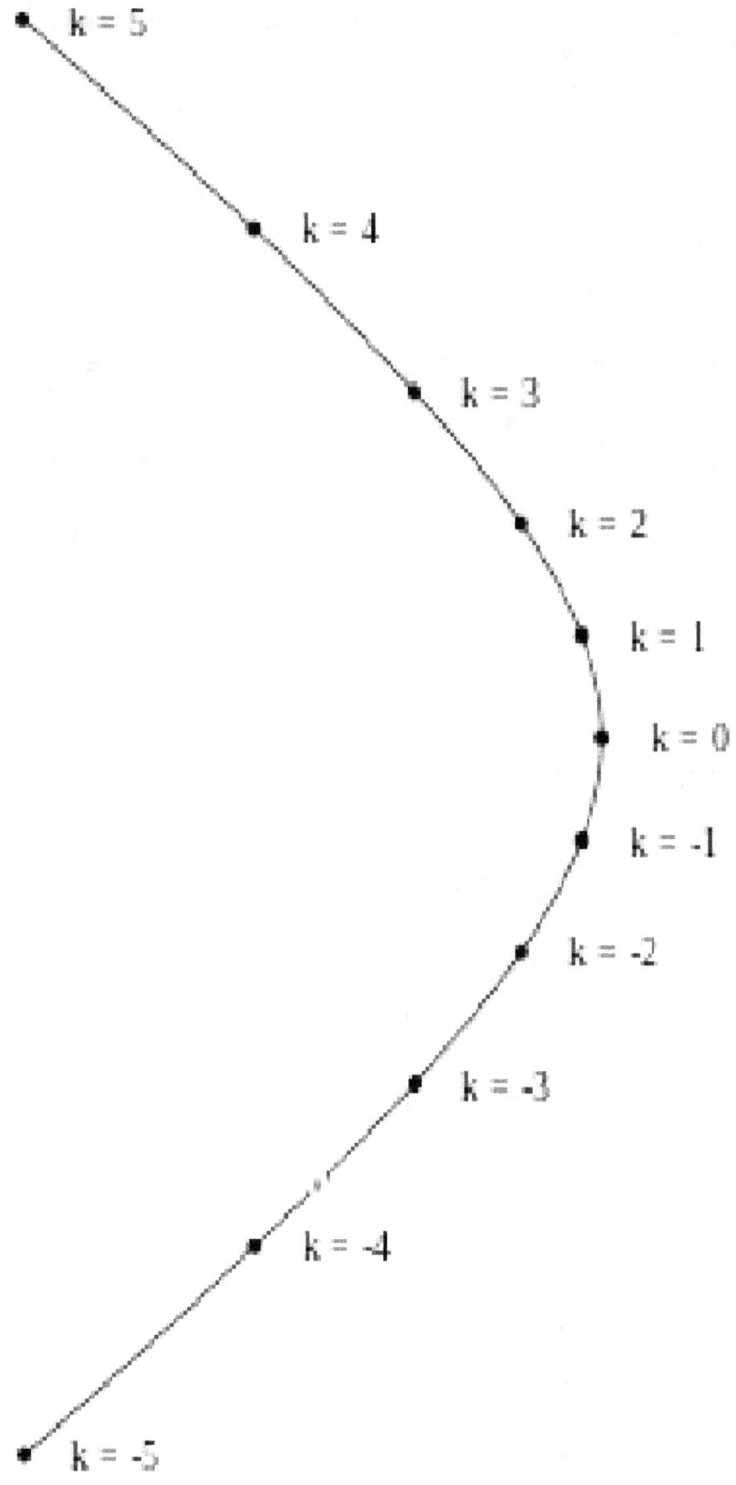

The left branch of an hyperbola parametrized using the hyperbolic cosinus and sinus function. The eccentricity is 1.2 and the arguments for the hyperbolic functions are 0.3 * k, k=-5,-4,..,5

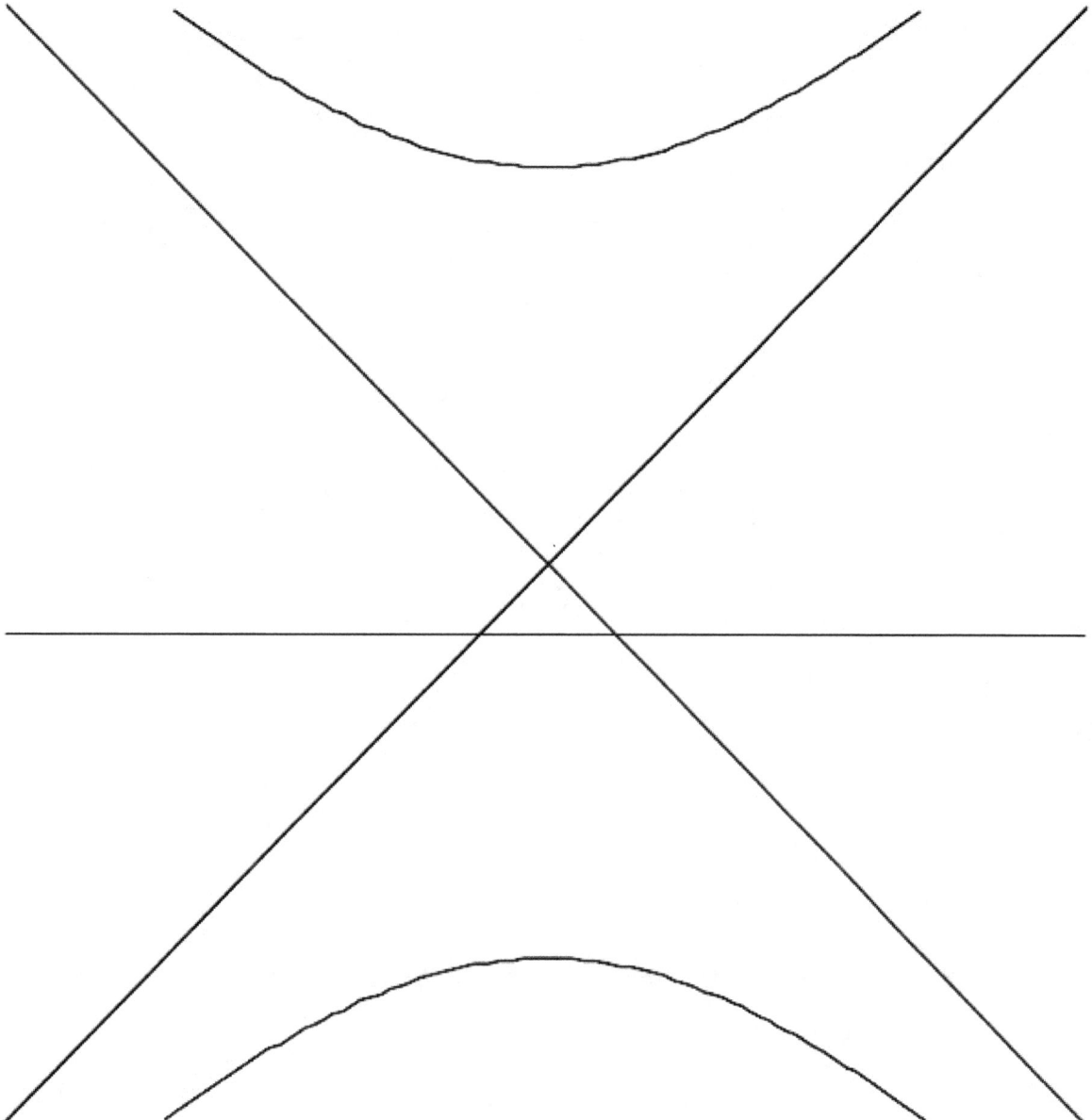

Figure 10: The hyperbola as a circle on the ground seen in perspective while gazing down slightly, showing the circle's (non-parallel) tangents as asymptotes. The portion above the horizon is normally invisible. When gazing straight ahead the tangents will be parallel and therefore intersect at the horizon instead of above, described further in the text.

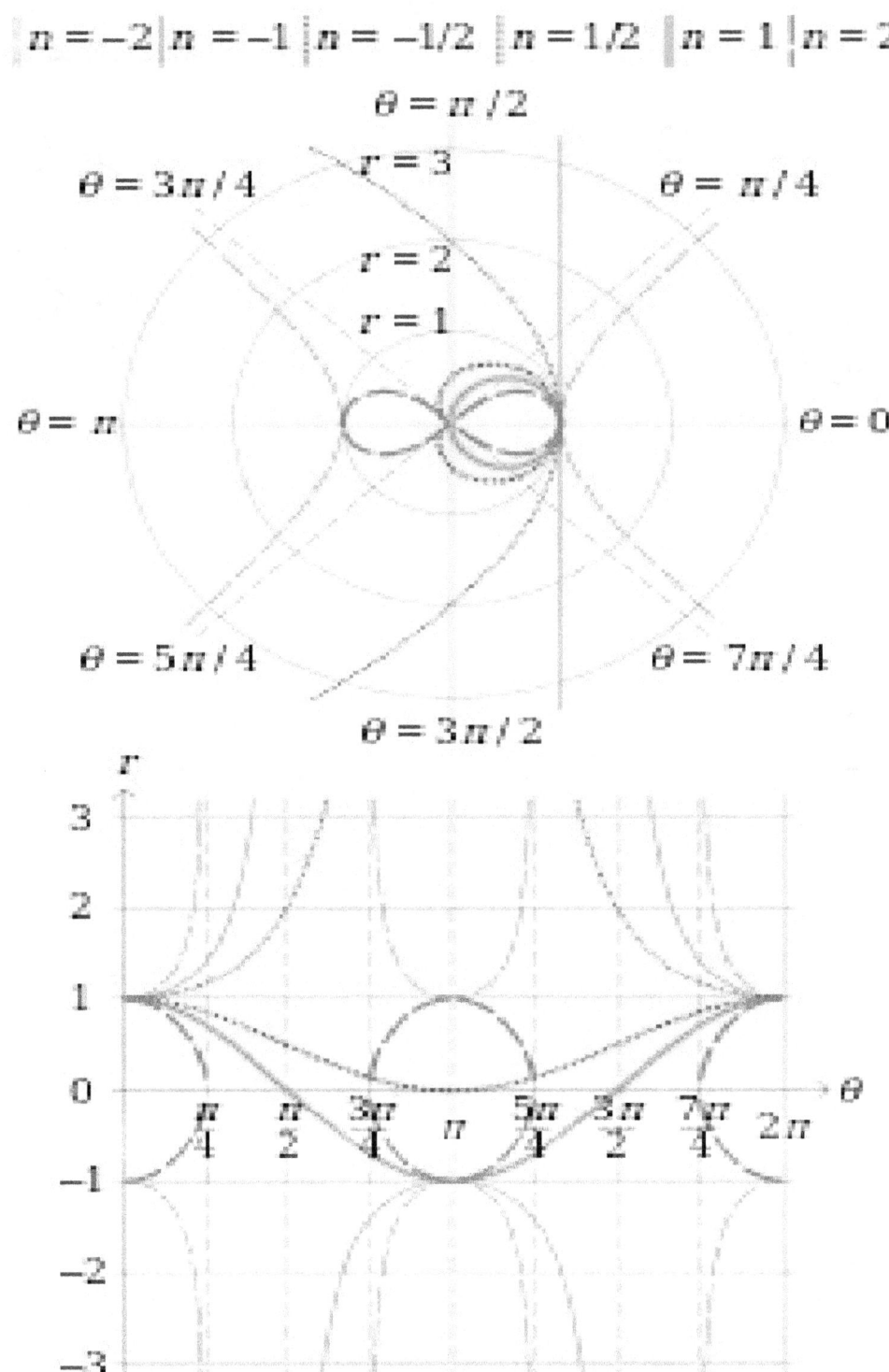

Comparison of sinusoidal spirals: equilateral hyperbola (n = -2), line (n = -1), parabola (n = -1/2), cardioid (n = 1/2), circle (n = 1) and lemniscate of Bernoulli (n = 2), where rn = 1n cos(nθ) in polar coordinates and their equivalents in rectangular coordinates.

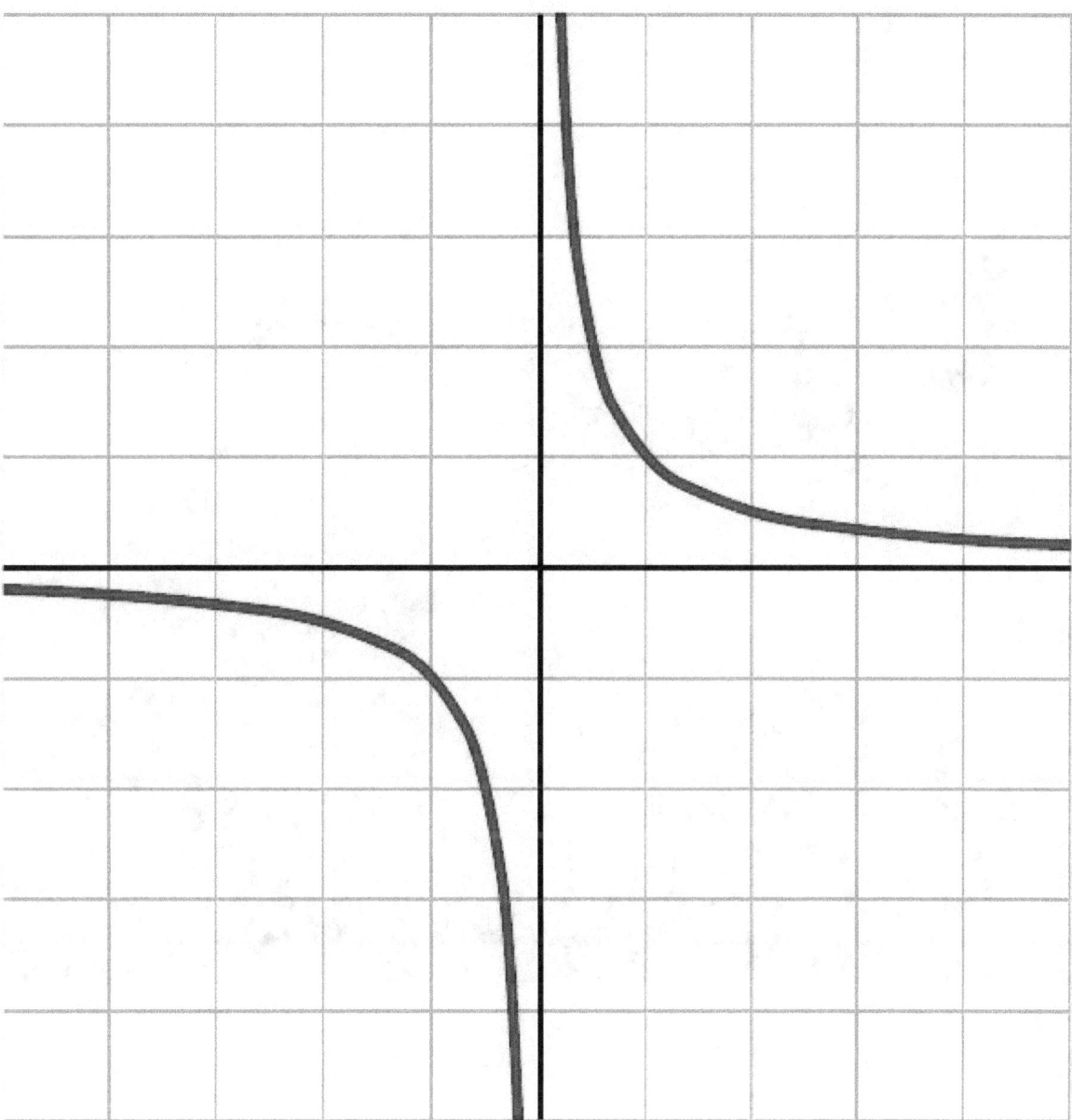

A graph of the rectangular hyperbola $y = \frac{1}{x}$, the reciprocal function

Hyperbolas as declination lines on a sundial

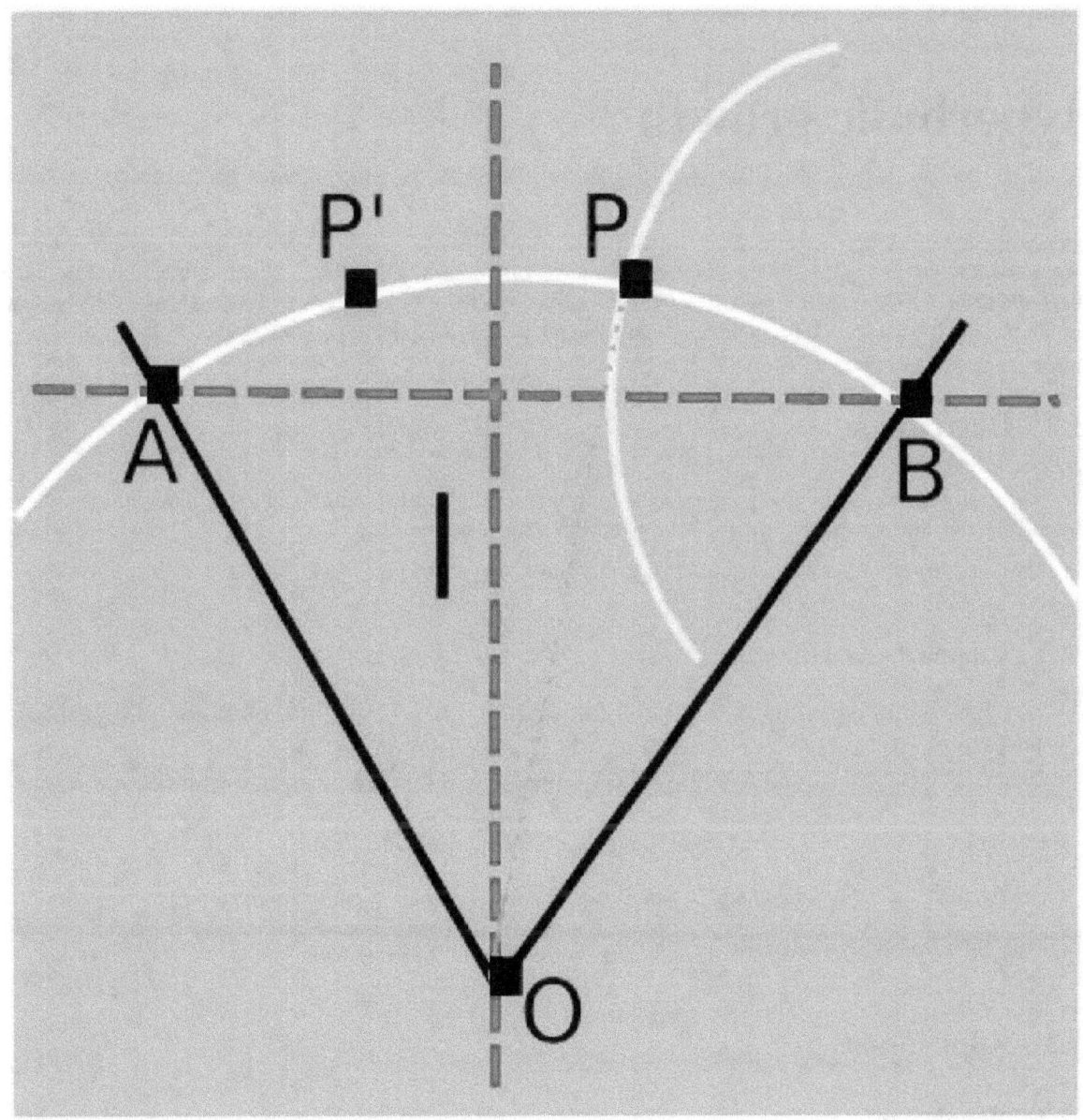

Trisecting an angle (AOB) using a hyperbola of eccentricity 2 (yellow curve)

Chapter 12

Hyperbolic growth

When a quantity grows towards a singularity under a finite variation (a "finite-time singularity") it is said to undergo **hyperbolic growth**.[1] More precisely, the reciprocal function $1/x$ has a hyperbola as a graph, and has a singularity at 0, meaning that the limit as $x \to 0$ is infinite: any similar graph is said to exhibit hyperbolic growth.

12.1 Description

If the output of a function is inversely proportional to its input, or inversely proportional to the difference from a given value x_0, the function will exhibit hyperbolic growth, with a singularity at x_0.

In the real world hyperbolic growth is created by certain non-linear positive feedback mechanisms.[2]

12.1.1 Comparisons with other growth

Like exponential growth and logistic growth, hyperbolic growth is highly nonlinear, but differs in important respects. These functions can be confused, as exponential growth, hyperbolic growth, and the first half of logistic growth are convex functions; however their asymptotic behavior (behavior as input gets large) differs dramatically:

- logistic growth is constrained (has a finite limit, even as time goes to infinity),

- exponential growth grows to infinity as time goes to infinity (but is always finite for finite time).

- hyperbolic growth has a singularity in finite time (grows to infinity at a finite time).

12.2 Applications

12.2.1 Population

Certain mathematical models suggest that until the early 1970s the world population underwent hyperbolic growth (see, e.g., *Introduction to Social Macrodynamics* by Andrey Korotayev *et al.*). It was also shown that until the 1970s the hyperbolic growth of the world population was accompanied by quadratic-hyperbolic growth of the world GDP, and developed a number of mathematical models describing both this phenomenon, and the World System withdrawal from the blow-up regime observed in the recent decades. The hyperbolic growth of the world population and quadratic-hyperbolic growth of the world GDP observed till the 1970s have been correlated by Andrey Korotayev and his colleagues to a non-linear second order positive feedback between the demographic growth and technological development, described by a chain of causation: technological growth leads to more carrying capacity of land for people, which leads to more people, which

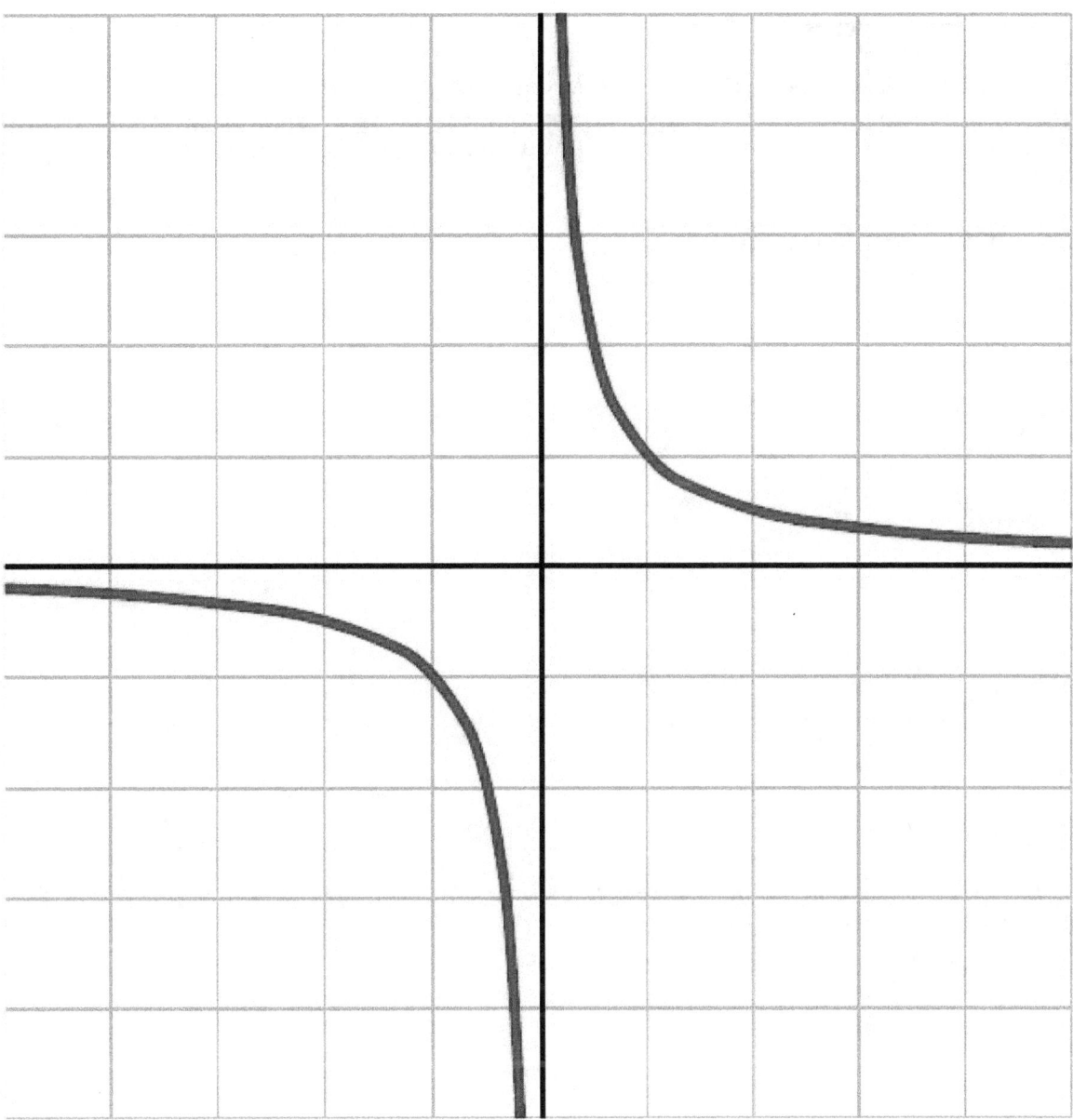

The reciprocal function, exhibiting hyperbolic growth.

leads to more inventors, which in turn leads to yet more technological growth, and on and on.[3] Other models suggest exponential growth, logistic growth, or other functions.

12.2.2   Queuing theory

Another example of hyperbolic growth can be found in queueing theory: the average waiting time of randomly arriving customers grows hyperbolically as a function of the average load ratio of the server. The singularity in this case occurs when the average amount of work arriving to the server equals the server's processing capacity. If the processing needs exceed the server's capacity, then there is no well-defined average waiting time, as the queue can grow without bound. A practical implication of this particular example is that for highly loaded queuing systems the average waiting time can be extremely sensitive to the processing capacity.

12.2.3 Enzyme kinetics

A further practical example of hyperbolic growth can be found in enzyme kinetics. When the rate of reaction (termed velocity) between an enzyme and substrate is plotted against various concentrations of the substrate, a hyperbolic plot is obtained for many simpler systems. When this happens, the enzyme is said to follow Michaelis-Menten kinetics.

12.3 Mathematical example

The function

$$x(t) = \frac{1}{t_c - t}$$

exhibits hyperbolic growth with a singularity at time t_c : in the limit as $t \to t_c$, the function goes to infinity.

More generally, the function

$$x(t) = \frac{K}{t_c - t}$$

exhibits hyperbolic growth, where K is a scale factor.

Note that this algebraic function can be regarded as analytical solution for the function's differential:[4]

$$\frac{dx}{dt} = \frac{K}{(t_c - t)^2} = \frac{x^2}{K}$$

This means that with hyperbolic growth the absolute growth rate of the variable x in the moment t is proportional to the square of the value of x in the moment t.

Respectively, the quadratic-hyperbolic function looks as follows:

$$x(t) = \frac{K}{(t_c - t)^2}.$$

12.4 See also

- Heinz von Foerster
- Technological singularity
- Paradigm shift
- Scientific mythology
- Social effect of evolutionary theory
- Deep ecology

12.4.1 Mathematics

- Mathematical singularity

12.4.2   Growth

- Exponential growth

- Logistic growth

12.5   References

- Alexander V. Markov, and Andrey V. Korotayev (2007). "Phanerozoic marine biodiversity follows a hyperbolic trend". Palaeoworld. Volume 16. Issue 4. Pages 311-318.

- Kremer, Michael. 1993. "Population Growth and Technological Change: One Million B.C. to 1990," The Quarterly Journal of Economics 108(3): 681-716.

- Korotayev A., Malkov A., Khaltourina D. 2006. *Introduction to Social Macrodynamics: Compact Macromodels of the World System Growth.* Moscow: URSS. ISBN 5-484-00414-4 .

- Rein Taagepera (1979) People, skills, and resources: An interaction model for world population growth. Technological Forecasting and Social Change 13, 13-30.

12.6   References

[1] See, e.g., Korotayev A., Malkov A., Khaltourina D. **Introduction to Social Macrodynamics: Compact Macromodels of the World System Growth.** Moscow: URSS Publishers, 2006. P. 19-20.

[2] See, e.g., Alexander V. Markov, and Andrey V. Korotayev (2007). "Phanerozoic marine biodiversity follows a hyperbolic trend". Palaeoworld. Volume 16. Issue 4. Pages 311-318.

[3] See, e.g., Korotayev A., Malkov A., Khaltourina D. **Introduction to Social Macrodynamics: Compact Macromodels of the World System Growth.** Moscow: URSS Publishers, 2006; Korotayev A. V. A Compact Macromodel of World System Evolution // Journal of World-Systems Research 11/1 (2005): 79–93.; for a detailed mathematical analysis of this issue see A Compact Mathematical Model of the World System Economic and Demographic Growth, 1 CE - 1973 CE.

[4] See, e.g., Korotayev A., Malkov A., Khaltourina D. **Introduction to Social Macrodynamics: Compact Macromodels of the World System Growth.** Moscow: URSS Publishers, 2006. P. 118-123.

Chapter 13

Multiplicative inverse

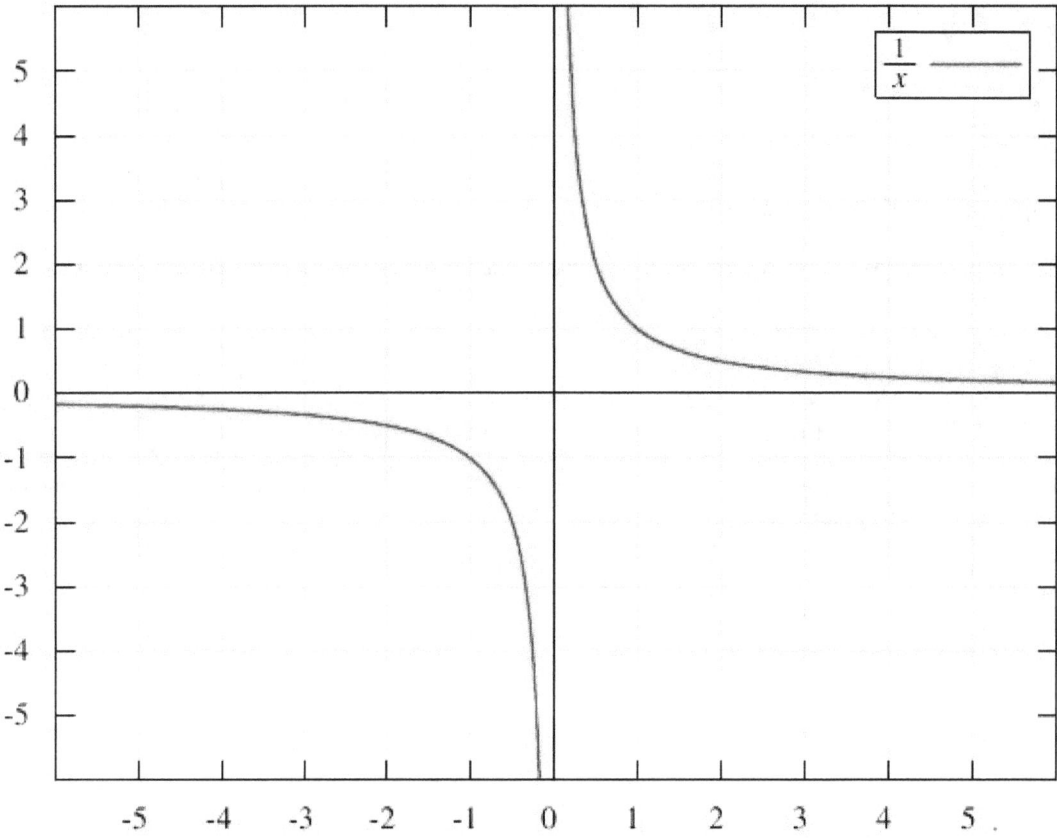

The reciprocal function: y = 1/x. For every x except 0, y represents its multiplicative inverse. The graph forms a rectangular hyperbola.

In mathematics, a **multiplicative inverse** or **reciprocal** for a number x, denoted by $1/x$ or x^{-1}, is a number which when multiplied by x yields the multiplicative identity, 1. The multiplicative inverse of a fraction a/b is b/a. For the multiplicative inverse of a real number, divide 1 by the number. For example, the reciprocal of 5 is one fifth (1/5 or 0.2), and the reciprocal of 0.25 is 1 divided by 0.25, or 4. The **reciprocal function**, the function $f(x)$ that maps x to $1/x$, is one of the simplest examples of a function which is its own inverse (an involution).

The term *reciprocal* was in common use at least as far back as the third edition of *Encyclopædia Britannica* (1797) to

119

describe two numbers whose product is 1; geometrical quantities in inverse proportion are described as *reciprocall* in a 1570 translation of Euclid's *Elements*.[1]

In the phrase *multiplicative inverse*, the qualifier *multiplicative* is often omitted and then tacitly understood (in contrast to the additive inverse). Multiplicative inverses can be defined over many mathematical domains as well as numbers. In these cases it can happen that $ab \neq ba$; then "inverse" typically implies that an element is both a left and right inverse.

The notation f^{-1} is sometimes also used for the inverse function of the function f, which is not in general equal to the multiplicative inverse. For example, the multiplicative inverse $1/(\sin x) = (\sin x)^{-1}$ is different from the inverse sin of x, denoted $\sin^{-1} x$ or arcsin x. Only for linear maps are they strongly related (see below). The terminology difference *reciprocal* versus *inverse* is not sufficient to make this distinction, since many authors prefer the opposite naming convention, probably for historical reasons (for example in French, the inverse function is preferably called application réciproque).

13.1   Examples and counterexamples

In the real numbers, zero does not have a reciprocal because no real number multiplied by 0 produces 1 (the product of any number with zero is zero). With the exception of zero, reciprocals of every real number are real, reciprocals of every rational number are rational, and reciprocals of every complex number are complex. The property that every element other than zero has a multiplicative inverse is part of the definition of a field, of which these are all examples. On the other hand, no integer other than 1 and -1 has an integer reciprocal, and so the integers are not a field.

In modular arithmetic, the modular multiplicative inverse of a is also defined: it is the number x such that $ax \equiv 1$ (mod n). This multiplicative inverse exists if and only if a and n are coprime. For example, the inverse of 3 modulo 11 is 4 because $4 \cdot 3 \equiv 1$ (mod 11). The extended Euclidean algorithm may be used to compute it.

The sedenions are an algebra in which every nonzero element has a multiplicative inverse, but which nonetheless has divisors of zero, i.e. nonzero elements x, y such that $xy = 0$.

A square matrix has an inverse if and only if its determinant has an inverse in the coefficient ring. The linear map that has the matrix A^{-1} with respect to some base is then the reciprocal function of the map having A as matrix in the same base. Thus, the two distinct notions of the inverse of a function are strongly related in this case, while they must be carefully distinguished in the general case (as noted above).

The trigonometric functions are related by the reciprocal identity: the cotangent is the reciprocal of the tangent; the secant is the reciprocal of the cosine; the cosecant is the reciprocal of the sine.

A ring in which every nonzero element has a multiplicative inverse is a division ring; likewise an algebra in which this holds is a division algebra.

13.2   Complex numbers

As mentioned above, the reciprocal of every nonzero complex number $z = a + bi$ is complex. It can be found by multiplying both top and bottom of $1/z$ by its complex conjugate $\bar{z} = a - bi$ and using the property that $z\bar{z} = \|z\|^2$, the absolute value of z squared, which is the real number $a^2 + b^2$:

$$\frac{1}{z} = \frac{\bar{z}}{z\bar{z}} = \frac{\bar{z}}{\|z\|^2} = \frac{a - bi}{a^2 + b^2} = \frac{a}{a^2 + b^2} - \frac{b}{a^2 + b^2}i.$$

In particular, if $\|z\|=1$ (z has unit magnitude), then $1/z = \bar{z}$. Consequently, the imaginary units, $\pm i$, have additive inverse equal to multiplicative inverse, and are the only complex numbers with this property. For example, additive and multiplicative inverses of i are $-(i) = -i$ and $1/i = -i$, respectively.

For a complex number in polar form $z = r(\cos \varphi + i \sin \varphi)$, the reciprocal simply takes the reciprocal of the magnitude and the negative of the angle:

$$\frac{1}{z} = \frac{1}{r}\left(\cos(-\varphi) + i\sin(-\varphi)\right).$$

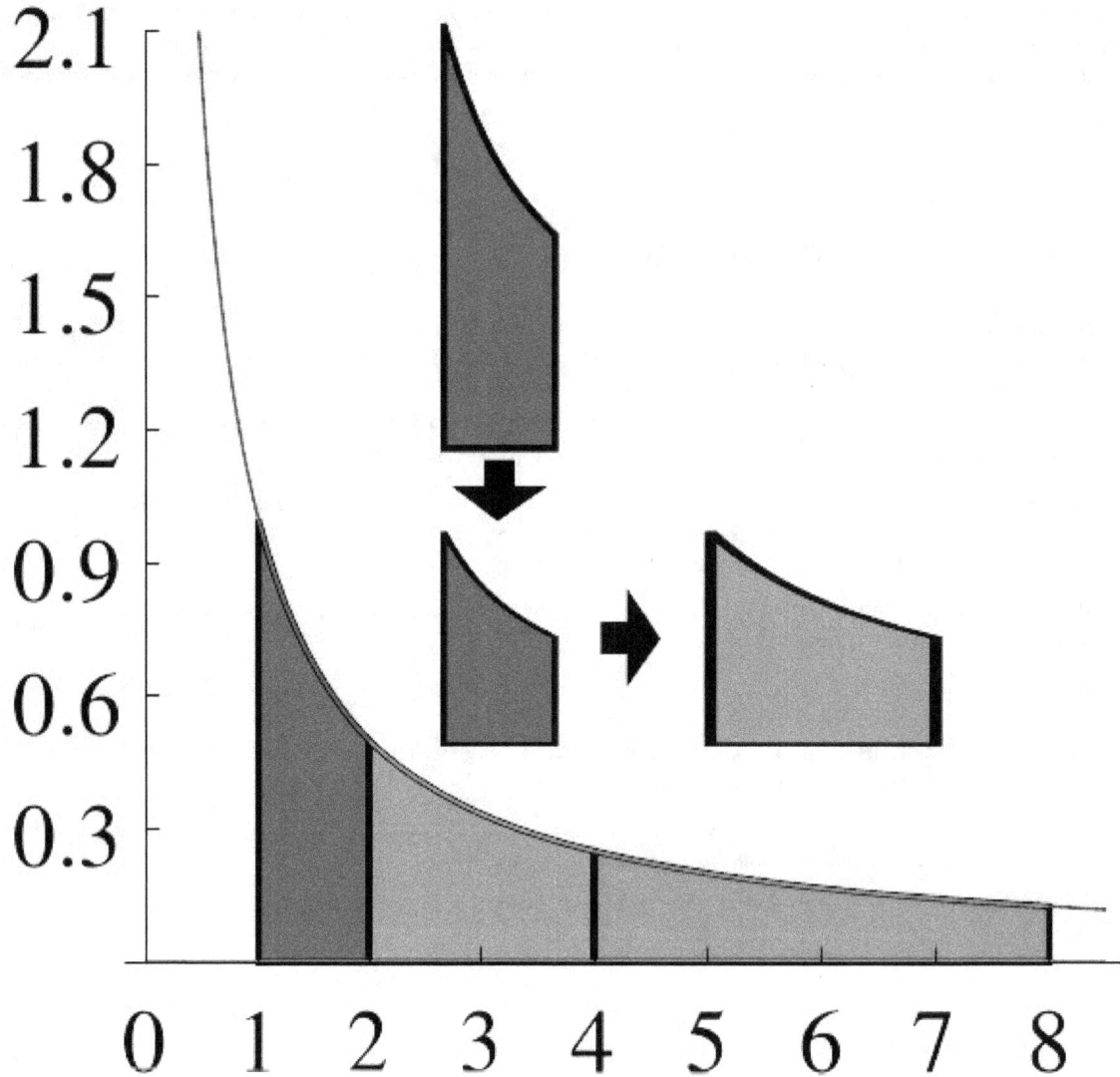

Geometric intuition for the integral of 1/x. The three integrals from 1 to 2, from 2 to 4, and from 4 to 8 are all equal. Each region is the previous region scaled vertically down by 50%, then horizontally by 200%. Extending this, the integral from 1 to 2^k is k times the integral from 1 to 2, just as ln 2^k = k ln 2.

13.3 Calculus

In real calculus, the derivative of $1/x = x^{-1}$ is given by the power rule with the power -1:

$$\frac{d}{dx}x^{-1} = (-1)x^{(-1)-1} = -x^{-2} = -\frac{1}{x^2}.$$

The power rule for integrals (Cavalieri's quadrature formula) cannot be used to compute the integral of $1/x$, because doing so would result in division by 0:

$$\int \frac{1}{x} dx = \frac{x^0}{0} + C$$

Instead the integral is given by:

$$\int_1^a \frac{1}{x} dx = \ln a,$$

$$\int \frac{1}{x} dx = \ln x + C.$$

where ln is the natural logarithm. To show this, note that $\frac{d}{dx} e^x = e^x$, so if $y = e^x$ and $x = \ln y$, we have:[2]

$$\frac{dy}{dx} = y \quad \Rightarrow \quad \frac{dy}{y} = dx \quad \Rightarrow \quad \int \frac{1}{y} dy = \int 1 \, dx \quad \Rightarrow \quad \int \frac{1}{y} dy = x + C = \ln y + C.$$

13.4   Algorithms

The reciprocal may be computed by hand with the use of long division.

Computing the reciprocal is important in many division algorithms, since the quotient a/b can be computed by first computing $1/b$ and then multiplying it by a. Noting that $f(x) = 1/x - b$ has a zero at $x = 1/b$, Newton's method can find that zero, starting with a guess x_0 and iterating using the rule:

$$x_{n+1} = x_n - \frac{f(x_n)}{f'(x_n)} = x_n - \frac{1/x_n - b}{-1/x_n^2} = 2x_n - bx_n^2 = x_n(2 - bx_n).$$

This continues until the desired precision is reached. For example, suppose we wish to compute $1/17 \approx 0.0588$ with 3 digits of precision. Taking $x_0 = 0.1$, the following sequence is produced:

$x_1 = 0.1(2 - 17 \times 0.1) = 0.03$

$x_2 = 0.03(2 - 17 \times 0.03) = 0.0447$

$x_3 = 0.0447(2 - 17 \times 0.0447) \approx 0.0554$

$x_4 = 0.0554(2 - 17 \times 0.0554) \approx 0.0586$

$x_5 = 0.0586(2 - 17 \times 0.0586) \approx 0.0588$

A typical initial guess can be found by rounding b to a nearby power of 2, then using bit shifts to compute its reciprocal.

In constructive mathematics, for a real number x to have a reciprocal, it is not sufficient that $x \neq 0$. There must instead be given a *rational* number r such that $0 < r < |x|$. In terms of the approximation algorithm described above, this is needed to prove that the change in y will eventually become arbitrarily small.

This iteration can also be generalised to a wider sort of inverses, e.g. matrix inverses.

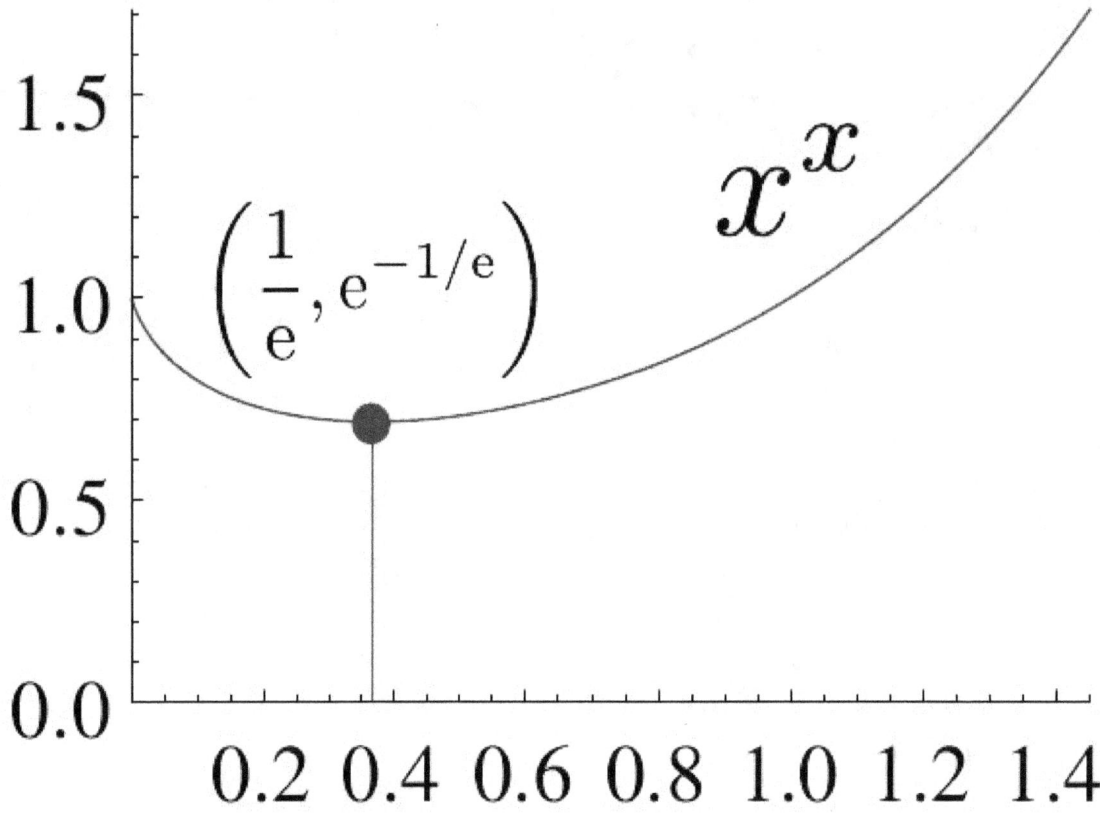

Graph of $f(x) = x^x$ showing the minimum at $(1/e, e^{-1/e})$.

13.5 Reciprocals of irrational numbers

Every number excluding zero has a reciprocal, and reciprocals of certain irrational numbers can have important special properties. Examples include the reciprocal of e (≈ 0.367879)and the golden ratio's reciprocal (≈ 0.618034). The first reciprocal is special because no other positive number can produce a lower number when put to the power of itself; $f(1/e)$ is the global minimum of $f(x) = x^x$. The second number is the only positive number that is equal to its reciprocal plus one: $\phi = 1/\phi + 1$. Its additive inverse is the only negative number that is equal to its reciprocal minus one: $-\phi = -1/\phi - 1$.

The function $f(n) = n + \sqrt{(n^2 + 1)}, n \in N, n > 0$ gives an infinite number of irrational numbers that differ with their reciprocal by an integer. For example, $f(2)$ is the irrational $2 + \sqrt{5}$. Its reciprocal $1/(2 + \sqrt{5})$ is $-2 + \sqrt{5}$, exactly 4 less. Such irrational numbers share a curious property: they have the same fractional part as their reciprocal.

13.6 Further remarks

If the multiplication is associative, an element x with a multiplicative inverse cannot be a zero divisor (meaning for some y, $xy = 0$ with neither x nor y equal to zero). To see this, it is sufficient to multiply the equation $xy = 0$ by the inverse of x (on the left), and then simplify using associativity. In the absence of associativity, the sedenions provide a counterexample.

The converse does not hold: an element which is not a zero divisor is not guaranteed to have a multiplicative inverse. Within **Z**, all integers except $-1, 0, 1$ provide examples; they are not zero divisors nor do they have inverses in **Z**. If the ring or algebra is finite, however, then all elements a which are not zero divisors do have a (left and right) inverse. For, first observe that the map $f(x) = ax$ must be injective: $f(x) = f(y)$ implies $x = y$:

$$ax = ay \quad \Rightarrow \quad ax - ay = 0$$
$$\Rightarrow \quad a(x - y) = 0$$
$$\Rightarrow \quad x - y = 0$$
$$\Rightarrow \quad x = y.$$

Distinct elements map to distinct elements, so the image consists of the same finite number of elements, and the map is necessarily surjective. Specifically, f (namely multiplication by a) must map some element x to 1, $ax = 1$, so that x is an inverse for a.

13.7   Applications

The expansion of the reciprocal $1/q$ in any base can also act [3] as a source of pseudo-random numbers, if q is a "suitable" safe prime, a prime of the form $2p + 1$ where p is also a prime. A sequence of pseudo-random numbers of length $q - 1$ will be produced by the expansion.

13.8   See also

- Division (mathematics)

- Fraction (mathematics)

- Group (mathematics)

- Ring (mathematics)

- Division algebra

- Exponential decay

- Unit fractions – reciprocals of integers

- Hyperbola

- Repeating decimal

- List of sums of reciprocals

13.9   Notes

[1] "In equall Parallelipipedons the bases are reciprokall to their altitudes". *OED* "Reciprocal" §3a. Sir Henry Billingsley translation of Elements XI, 34.

[2] Anthony, Dr. "Proof that INT(1/x)dx = lnx". *Ask Dr. Math*. Drexel University. Retrieved 22 March 2013.

[3] Mitchell, Douglas W.. "A nonlinear random number generator with known, long cycle length." *Cryptologia* 17, January 1993, 55-62.

13.10   References

- Maximally Periodic Reciprocals, Matthews R.A.J. *Bulletin of the Institute of Mathematics and its Applications* vol 28 pp 147–148 1992

Chapter 14

Linear function

In mathematics, the term **linear function** refers to two distinct, although related, notions:[1]

- In calculus and related areas, a linear function is a polynomial function of degree zero or one, or is the zero polynomial.[2]

- In linear algebra and functional analysis, a linear function is a linear map.[3]

14.1   As a polynomial function

Main article: Linear function (calculus)

In calculus, analytic geometry and related areas, a linear function is a polynomial of degree one or less, including the zero polynomial (the latter not being considered to have degree zero).

When the function is of only one variable, it is of the form

$$f(x) = ax + b,$$

where a and b are constants, often real numbers. The graph of such a function of one variable is a nonvertical line. a is frequently referred to as the slope of the line, and b as the intercept.

For a function $f(x_1, \ldots, x_k)$ of any finite number of independent variables, the general formula is

$$f(x_1, \ldots, x_k) = b + a_1 x_1 + \ldots + a_k x_k$$

and the graph is a hyperplane of dimension k.

A constant function is also considered linear in this context, as it is a polynomial of degree zero or is the zero polynomial. Its graph, when there is only one independent variable, is a horizontal line.

In this context, the other meaning (a linear map) may be referred to as a homogeneous linear function or a linear form. In the context of linear algebra, this meaning (polynomial functions of degree 0 or 1) is a special kind of affine map.

14.2   As a linear map

Main article: Linear map

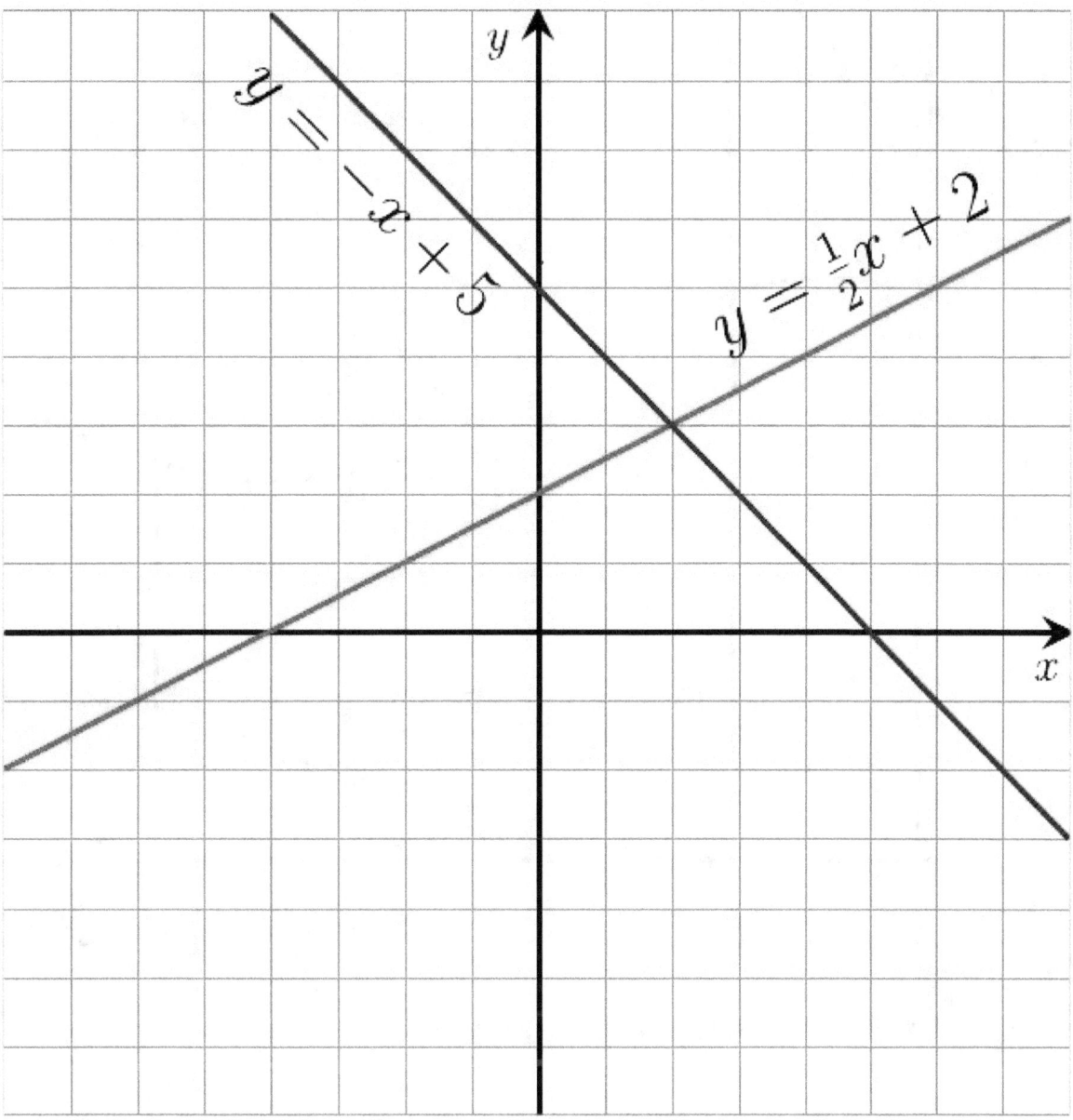

Graphs of two linear (polynomial) functions.

In linear algebra, a linear function is a map f between two vector spaces that preserves vector addition and scalar multiplication:

$$f(\mathbf{x} + \mathbf{y}) = f(\mathbf{x}) + f(\mathbf{y})$$

$$f(a\mathbf{x}) = a f(\mathbf{x}).$$

Here a denotes a constant belonging to some field K of scalars (for example, the real numbers) and $\mathbf{x}$ and $\mathbf{y}$ are elements of a vector space, which might be K itself.

Some authors use "linear function" only for linear maps that take values in the scalar field;[4] these are also called linear functionals.

The "linear functions" of calculus qualify as "linear maps" when (and only when) $f(0[,\ldots,0]) = 0$, or, equivalently, when the constant $b = 0$. Geometrically, the graph of the function must pass through the origin.

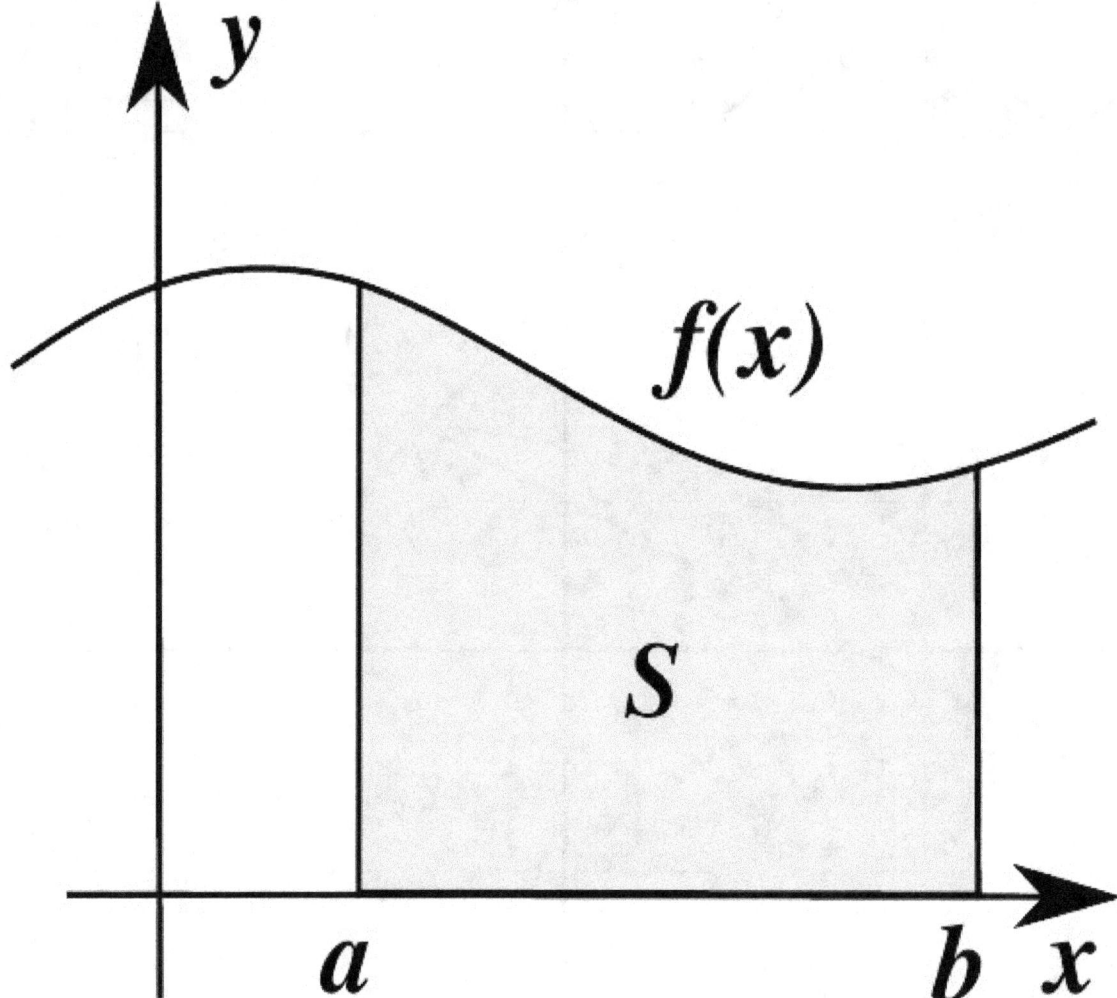

The integral of a function is a linear map from the vector space of integrable functions to the real numbers.

14.3   See also

- Homogeneous function

- Nonlinear system

- Piecewise linear function

- Linear interpolation

- Discontinuous linear map

14.4   Notes

[1] "The term *linear function*, which is not used here, means a linear form in some textbooks and an affine function in others." Vaserstein 2006, p. 50-1

[2] Stewart 2012, p. 23

[3] Shores 2007, p. 71

[4] Gelfand 1961

14.5   References

- Izrail Moiseevich Gelfand (1961), *Lectures on Linear Algebra*, Interscience Publishers, Inc., New York. Reprinted by Dover, 1989. ISBN 0-486-66082-6

- Thomas S. Shores (2007), *Applied Linear Algebra and Matrix Analysis*, Undergraduate Texts in Mathematics, Springer. ISBN 0-387-33195-6

- James Stewart (2012), *Calculus: Early Transcendentals*, edition 7E, Brooks/Cole. ISBN 978-0-538-49790-9

- Leonid N. Vaserstein (2006), "Linear Programming", in Leslie Hogben, ed., *Handbook of Linear Algebra*, Discrete Mathematics and Its Applications, Chapman and Hall/CRC, chap. 50. ISBN 1-584-88510-6

14.6   External links

Chapter 15

Exponential function

This article is about the natural exponential function e^x. For the general exponential of the form b^x and any base b, see Exponentiation. For exponentially increasing functions of the form cb^x, see Exponential growth.
 In mathematics, an exponential function is a function of the form

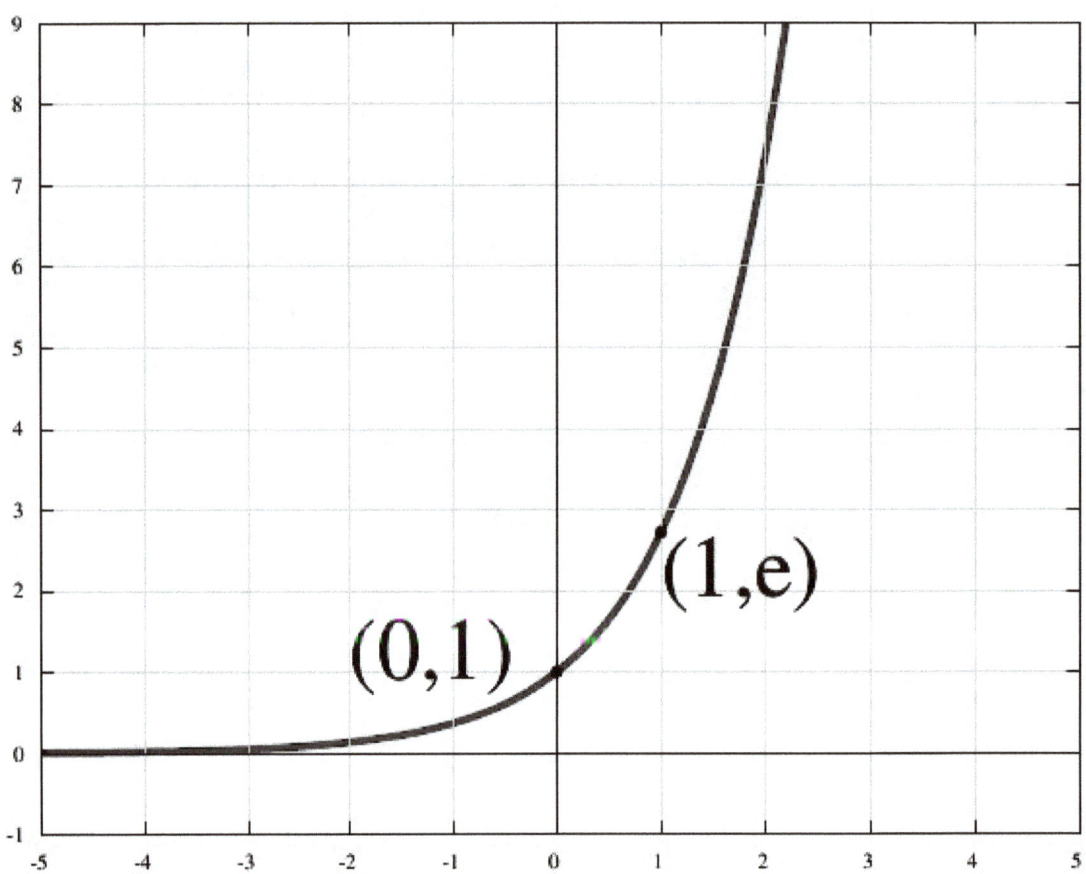

The natural exponential function $y = e^x$

$f(x) = b^x.$

The input variable x occurs as an exponent – hence the name. A function of the form $f(x) = b^{x \pm c}$ is also considered an exponential function, and a function of the form $f(x) = a \cdot b^x$ can be re-written as $f(x) = b^{x \pm c}$ by the use of logarithms and so is an exponential function.

In contexts where the base b is not specified, especially in more theoretical contexts, the term **exponential function** is almost always understood to mean the **natural exponential function**

$$x \mapsto e^x,$$

where e is Euler's number, a number (approximately 2.718281828). The reason this number e is considered the "natural" base of exponential functions is that this function is its own derivative.[1][2] Every exponential function is directly proportional to its own derivative, but only when the base is e does the constant of proportionality equal 1.

The exponential function is used to model a relationship in which a constant change in the independent variable gives the same proportional change (i.e. percentage increase or decrease) in the dependent variable. The function is often written as $\exp(x)$, especially when it is impractical to write the independent variable as a superscript. The exponential function is widely used in physics, chemistry, engineering, mathematical biology, economics and mathematics.

The graph of $y = e^x$ is upward-sloping, and increases faster as x increases. The graph always lies above the x-axis but can get arbitrarily close to it for negative x; thus, the x-axis is a horizontal asymptote. The slope of the tangent to the graph at each point is equal to its y coordinate at that point. The inverse function is the natural logarithm $\ln(x)$; because of this, some old texts[3] refer to the exponential function as the **antilogarithm**.

In general, the variable x can be any real or complex number or even an entirely different kind of mathematical object; see the formal definition below.

15.1   Formal definition

Main article: Characterizations of the exponential function

The exponential function e^x can be characterized in a variety of equivalent ways. In particular it may be defined by the following power series:[4]

$$e^x = \sum_{n=0}^{\infty} \frac{x^n}{n!} = 1 + x + \frac{x^2}{2!} + \frac{x^3}{3!} + \frac{x^4}{4!} + \cdots$$

Using an alternate definition for the exponential function leads to the same result when expanded as a Taylor series.

Less commonly, e^x is defined as the solution y to the equation

$$x = \int_1^y \frac{dt}{t}$$

It is also the following limit:[5]

$$e^x = \lim_{n \to \infty} \left(1 + \frac{x}{n}\right)^n$$

15.2   Overview

The exponential function arises whenever a quantity grows or decays at a rate proportional to its current value. One such situation is continuously compounded interest, and in fact it was this that led Jacob Bernoulli in 1683[6] to the number

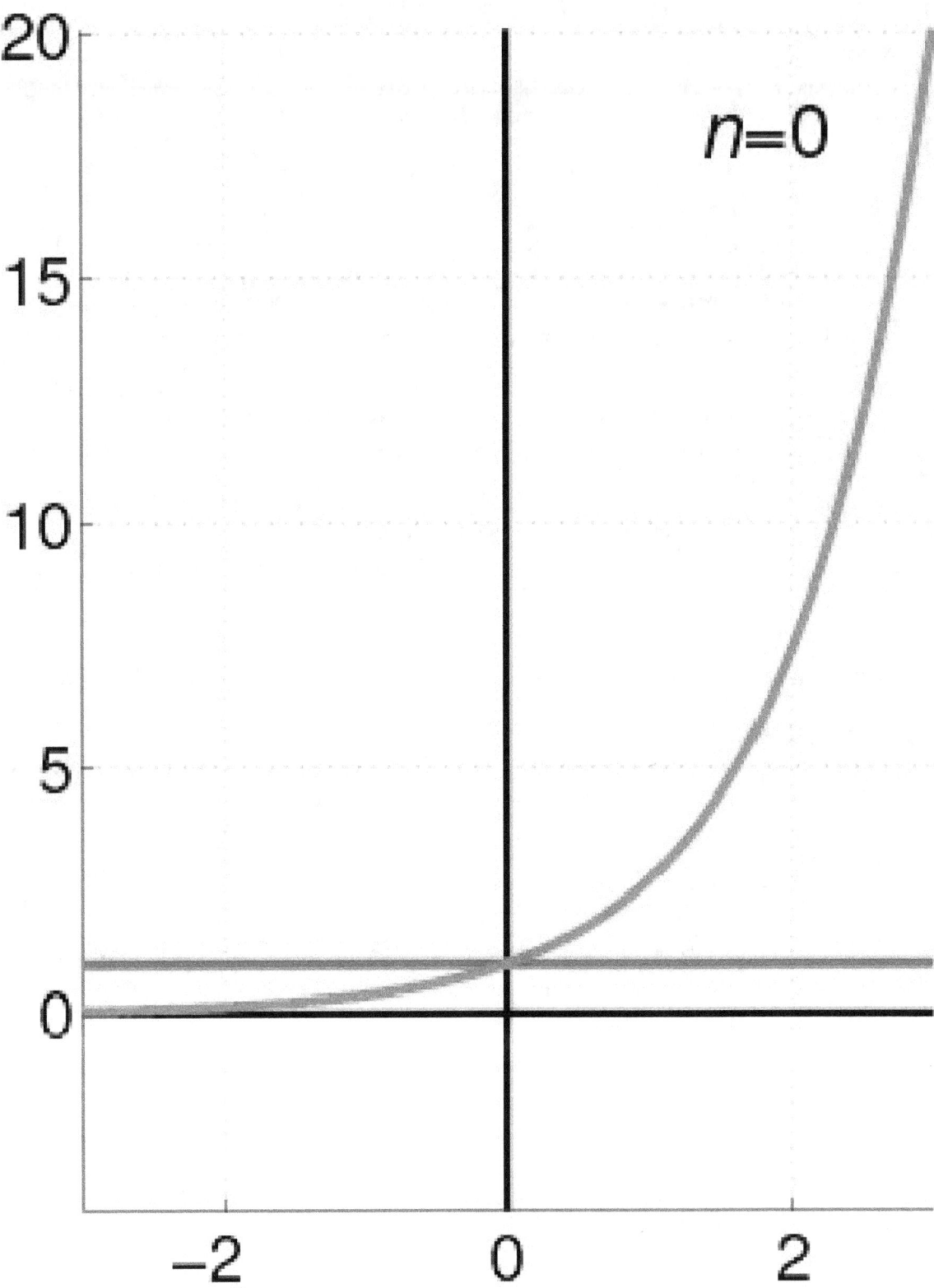

The exponential function (in blue), and the sum of the first n + 1 terms of the power series on the left (in red).

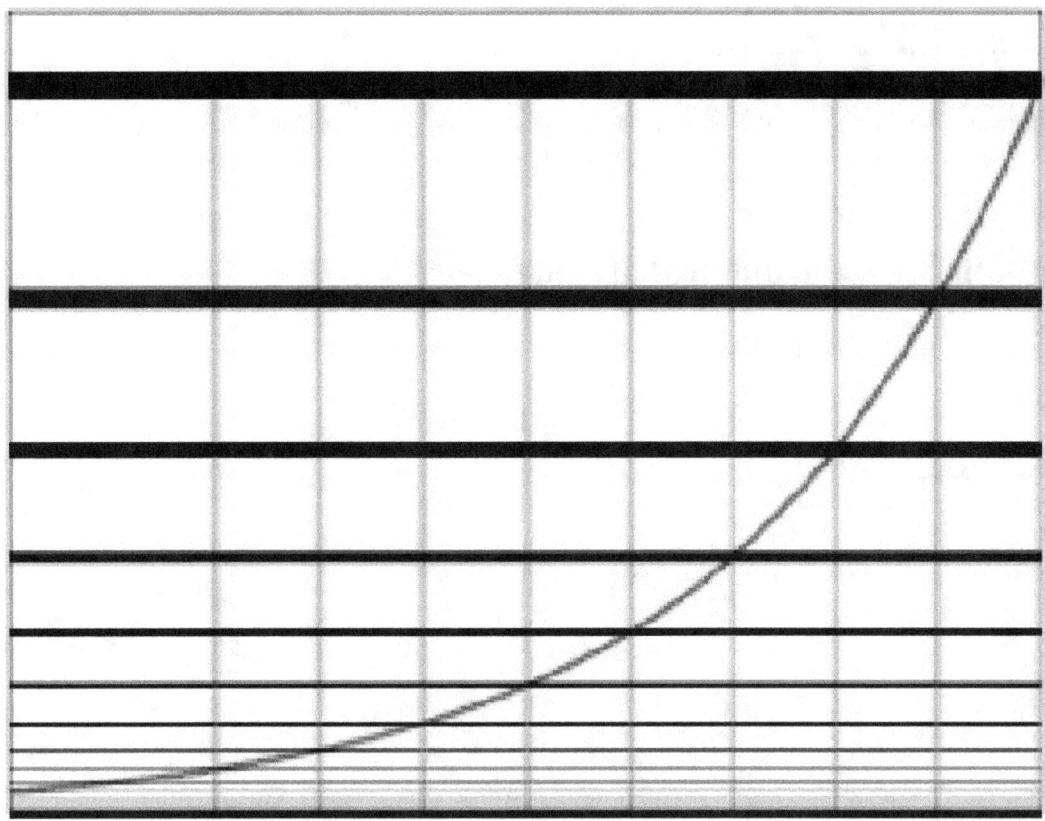

The red curve is the exponential function. The black horizontal lines show where it crosses the green vertical lines.

$$\lim_{n \to \infty} \left(1 + \frac{1}{n}\right)^n$$

now known as e. Later, in 1697, Johann Bernoulli studied the calculus of the exponential function.[6]

If a principal amount of 1 earns interest at an annual rate of x compounded monthly, then the interest earned each month is $x/12$ times the current value, so each month the total value is multiplied by $(1+x/12)$, and the value at the end of the year is $(1+x/12)^{12}$. If instead interest is compounded daily, this becomes $(1+x/365)^{365}$. Letting the number of time intervals per year grow without bound leads to the limit definition of the exponential function,

$$\exp(x) = \lim_{n \to \infty} \left(1 + \frac{x}{n}\right)^n$$

first given by Euler.[5] This is one of a number of characterizations of the exponential function; others involve series or differential equations.

From any of these definitions it can be shown that the exponential function obeys the basic exponentiation identity,

$$\exp(x + y) = \exp(x) \cdot \exp(y)$$

which is why it can be written as e^x.

The derivative (rate of change) of the exponential function is the exponential function itself. More generally, a function with a rate of change *proportional* to the function itself (rather than equal to it) is expressible in terms of the exponential function. This function property leads to exponential growth and exponential decay.

The exponential function extends to an entire function on the complex plane. Euler's formula relates its values at purely imaginary arguments to trigonometric functions. The exponential function also has analogues for which the argument is a matrix, or even an element of a Banach algebra or a Lie algebra.

15.3 Derivatives and differential equations

The importance of the exponential function in mathematics and the sciences stems mainly from properties of its derivative. In particular,

$$\frac{d}{dx}e^x = e^x$$

Proof:

$$e^x = 1 + x + \frac{x^2}{2!} + \frac{x^3}{3!} + \frac{x^4}{4!} + \frac{x^5}{5!} + \cdots$$

$$\frac{d}{dx}e^x = \frac{d}{dx}\left(1 + x + \frac{x^2}{2!} + \frac{x^3}{3!} + \frac{x^4}{4!} + \frac{x^5}{5!} + \cdots\right)$$

$$= 0 + 1 + \frac{2x}{2!} + \frac{3x^2}{3!} + \frac{4x^3}{4!} + \frac{5x^4}{5!} + \cdots$$

$$= 1 + x + \frac{x^2}{2!} + \frac{x^3}{3!} + \frac{x^4}{4!} + \frac{x^5}{5!} + \cdots$$

$$= e^x$$

That is, e^x is its own derivative and hence is a simple example of a Pfaffian function. Functions of the form ce^x for constant c are the only functions with that property (by the Picard–Lindelöf theorem). Other ways of saying the same thing include:

- The slope of the graph at any point is the height of the function at that point.

- The rate of increase of the function at x is equal to the value of the function at x.

- The function solves the differential equation $y' = y$.

- exp is a fixed point of derivative as a functional.

If a variable's growth or decay rate is proportional to its size—as is the case in unlimited population growth (see Malthusian catastrophe), continuously compounded interest, or radioactive decay—then the variable can be written as a constant times an exponential function of time. Explicitly for any real constant k, a function $f: \mathbf{R} \rightarrow \mathbf{R}$ satisfies $f' = kf$ if and only if $f(x) = ce^{kx}$ for some constant c.

Furthermore for any differentiable function $f(x)$, we find, by the chain rule:

$$\frac{d}{dx}e^{f(x)} = f'(x)e^{f(x)}$$

15.4 Continued fractions for e^x

A continued fraction for e^x can be obtained via an identity of Euler:

$$e^x = 1 + \cfrac{x}{1 - \cfrac{x}{x+2 - \cfrac{2x}{x+3 - \cfrac{3x}{x+4 - \ddots}}}}$$

The following generalized continued fraction for e^z converges more quickly:[7]

$$e^z = 1 + \cfrac{2z}{2 - z + \cfrac{z^2}{6 + \cfrac{z^2}{10 + \cfrac{z^2}{14 + \ddots}}}}$$

or, by applying the substitution $z = x/y$:

$$e^{\frac{x}{y}} = 1 + \cfrac{2x}{2y - x + \cfrac{x^2}{6y + \cfrac{x^2}{10y + \cfrac{x^2}{14y + \ddots}}}}$$

with a special case for $z = 2$:

$$e^2 = 1 + \cfrac{4}{0 + \cfrac{2^2}{6 + \cfrac{2^2}{10 + \cfrac{2^2}{14 + \ddots}}}} = 7 + \cfrac{2}{5 + \cfrac{1}{7 + \cfrac{1}{9 + \cfrac{1}{11 + \ddots}}}}$$

This formula also converges, though more slowly, for $z > 2$. For example:

$$e^3 = 1 + \cfrac{6}{-1 + \cfrac{3^2}{6 + \cfrac{3^2}{10 + \cfrac{3^2}{14 + \ddots}}}} = 13 + \cfrac{54}{7 + \cfrac{9}{14 + \cfrac{9}{18 + \cfrac{9}{22 + \ddots}}}}$$

15.5 Complex plane

As in the real case, the exponential function can be defined on the complex plane in several equivalent forms. One such definition parallels the power series definition for real numbers, where the real variable is replaced by a complex one:

$$e^z = \sum_{n=0}^{\infty} \frac{z^n}{n!}$$

The exponential function is periodic with imaginary period $2\pi i$ and can be written as

$$e^{a+bi} = e^a (\cos b + i \sin b)$$

where a and b are real values and on the right the real functions must be used if used as a definition[8] (see also Euler's formula). This formula connects the exponential function with the trigonometric functions and to the hyperbolic functions.

When considered as a function defined on the complex plane, the exponential function retains the properties

- $e^{z+w} = e^z e^w$
- $e^0 = 1$
- $e^z \neq 0$
- $\frac{d}{dz} e^z = e^z$
- $(e^z)^n = e^{nz}, n \in \mathbb{Z}$

for all z and w.

The exponential function is an entire function as it is holomorphic over the whole complex plane. It takes on every complex number excepting 0 as value; that is, 0 is a lacunary value of the exponential function. This is an example of Picard's little theorem that any non-constant entire function takes on every complex number as value with at most one value excepted.

Extending the natural logarithm to complex arguments yields the complex logarithm $\log z$, which is a multivalued function.

We can then define a more general exponentiation:

$$z^w = e^{w \log z}$$

for all complex numbers z and w. This is also a multivalued function, even when z is real. This distinction is problematic, as the multivalued functions $\log z$ and z^w are easily confused with their single-valued equivalents when substituting a real number for z. The rule about multiplying exponents for the case of positive real numbers must be modified in a multivalued context:

$$(e^z)^w \neq e^{zw} \text{, but rather } (e^z)^w = e^{(z+2\pi in)w} \text{ multivalued over integers } n$$

See failure of power and logarithm identities for more about problems with combining powers.

The exponential function maps any line in the complex plane to a logarithmic spiral in the complex plane with the center at the origin. Two special cases might be noted: when the original line is parallel to the real axis, the resulting spiral never closes in on itself; when the original line is parallel to the imaginary axis, the resulting spiral is a circle of some radius.

- Plots of the exponential function on the complex plane
- $z = \text{Re}(e^{x+iy})$
- $z = \text{Im}(e^{x+iy})$
- $z = |e^{x+iy}|$

15.5.1   Computation of a^b where both a and b are complex

Main article: Exponentiation

Complex exponentiation a^b can be defined by converting a to polar coordinates and using the identity $(e^{\ln(a)})^b = a^b$:

$$a^b = (re^{\theta i})^b = (e^{\ln(r)+\theta i})^b = e^{(\ln(r)+\theta i)b}$$

However, when b is not an integer, this function is multivalued, because θ is not unique (see failure of power and logarithm identities).

15.6   Matrices and Banach algebras

The power series definition of the exponential function makes sense for square matrices (for which the function is called the matrix exponential) and more generally in any Banach algebra B. In this setting, $e^0 = 1$, and e^x is invertible with inverse e^{-x} for any x in B. If $xy = yx$, then $e^{x+y} = e^x e^y$, but this identity can fail for noncommuting x and y.

Some alternative definitions lead to the same function. For instance, e^x can be defined as $\lim_{n \to \infty} \left(1 + \frac{x}{n}\right)^n$

Or e^x can be defined as $f(1)$, where $f : \mathbf{R} \to B$ is the solution to the differential equation $f'(t) = xf(t)$ with initial condition $f(0) = 1$.

15.7   Lie algebras

Given a Lie group G and its associated Lie algebra $\mathfrak{g}$, the exponential map is a map $\mathfrak{g} \to G$ satisfying similar properties. In fact, since $\mathbf{R}$ is the Lie algebra of the Lie group of all positive real numbers under multiplication, the ordinary exponential function for real arguments is a special case of the Lie algebra situation. Similarly, since the Lie group GL(n,$\mathbf{R}$) of invertible $n \times n$ matrices has as Lie algebra M(n,$\mathbf{R}$), the space of all $n \times n$ matrices, the exponential function for square matrices is a special case of the Lie algebra exponential map.

The identity $\exp(x + y) = \exp(x)\exp(y)$ can fail for Lie algebra elements x and y that do not commute; the Baker–Campbell–Hausdorff formula supplies the necessary correction terms.

15.8   Double exponential function

Main article: double exponential function

The term *double exponential function* can have two meanings:

* a function with two exponential terms, each with a different exponent such as in $e^{3x} - e^{4x-2}$

* a function $f(x) = a^{a^x}$; this grows even faster than an exponential function; for example, if $a = 10$, then $f(-1) = 1.26$, $f(0) = 10$, $f(1) = 10^{10}$, $f(2) = 10^{100}$ = googol, ..., $f(100)$ = googolplex.

Factorials grow faster than exponential functions, but slower than double-exponential functions. Fermat numbers, generated by $F(m) = 2^{2^m} + 1$ and double Mersenne numbers generated by $MM(p) = 2^{2^p-1} - 1$ are examples of double exponential functions.

15.9 Similar properties of *e* and the function *e*^{z}

The function e^z is not in $\mathbf{C}(z)$ (i.e., is not the quotient of two polynomials with complex coefficients).

For *n* distinct complex numbers $\{a_1, ..., an\}$, the set $\{e^{a_1 z}, ..., e^{anz}\}$ is linearly independent over $\mathbf{C}(z)$.

The function e^z is transcendental over $\mathbf{C}(z)$.

15.10 See also

- Approximating Natural exponents (log base e)

- Carlitz exponential, a characteristic *p* analogue

- Characterizations of the exponential function

- *e* (mathematical constant)

- Exponential decay

- Exponential field

- Exponential growth

- Exponentiation

- Half-exponential function – a compositional square root of an exponential function

- List of exponential topics

- List of integrals of exponential functions

- *p*-adic exponential function

- Padé approximation – it can be used to approximate the exponential function by a fraction of polynomial functions

- Tetration

15.11 References

[1] Goldstein, Lay, Schneider, Asmar, *Brief calculus and its applications*, 11th ed., Prentice–Hall, 2006.

[2] *"This natural exponential function is identical with its derivative. This is really the source of all the properties of the exponential function, and the basic reason for its importance in applications..."* – p.448 of Courant and Robbins, *What is mathematics? An elementary approach to ideas and methods* (edited by Stewart), 2nd revised edition, Oxford Univ. Press, 1996.

[3] "Inverse Use of a Table of Logarithms; that is, given a logarithm, to find the number corresponding to it, (called its antilogarithm)..." – p.12 of Converse and Durrell, *Plane and spherical trigonometry*, C.E. Merrill co., 1911.

[4] Rudin, Walter (1987). *Real and complex analysis* (3rd ed.). New York: McGraw-Hill. p. 1. ISBN 978-0-07-054234-1.

[5] Eli Maor, *e: the Story of a Number*, p.156.

[6] John J O'Connor; Edmund F Robertson. "The number e". *School of Mathematics and Statistics*. University of St Andrews, Scotland. Retrieved 2011-06-13.

[7] "A.2.2 The exponential function." L. Lorentzen and H. Waadeland, *Continued Fractions*, Atlantis Studies in Mathematics, page 268.

[8] Ahlfors, Lars V. (1953). *Complex analysis*. McGraw–Hill Book Company, Inc.

15.12 External links

- Hazewinkel, Michiel, ed. (2001), "Exponential function", *Encyclopedia of Mathematics*, Springer, ISBN 978-1-55608-010-4
- Complex exponential function at PlanetMath.org.
- Derivative of exponential function at PlanetMath.org.
- Derivative of exponential function interactive graph
- Weisstein, Eric W., "Exponential Function", *MathWorld*.
- Taylor Series Expansions of Exponential Functions at efunda.com
- Complex exponential interactive graphic
- Derivative of $\exp(x^n)$ by limit definition
- General exponential limit

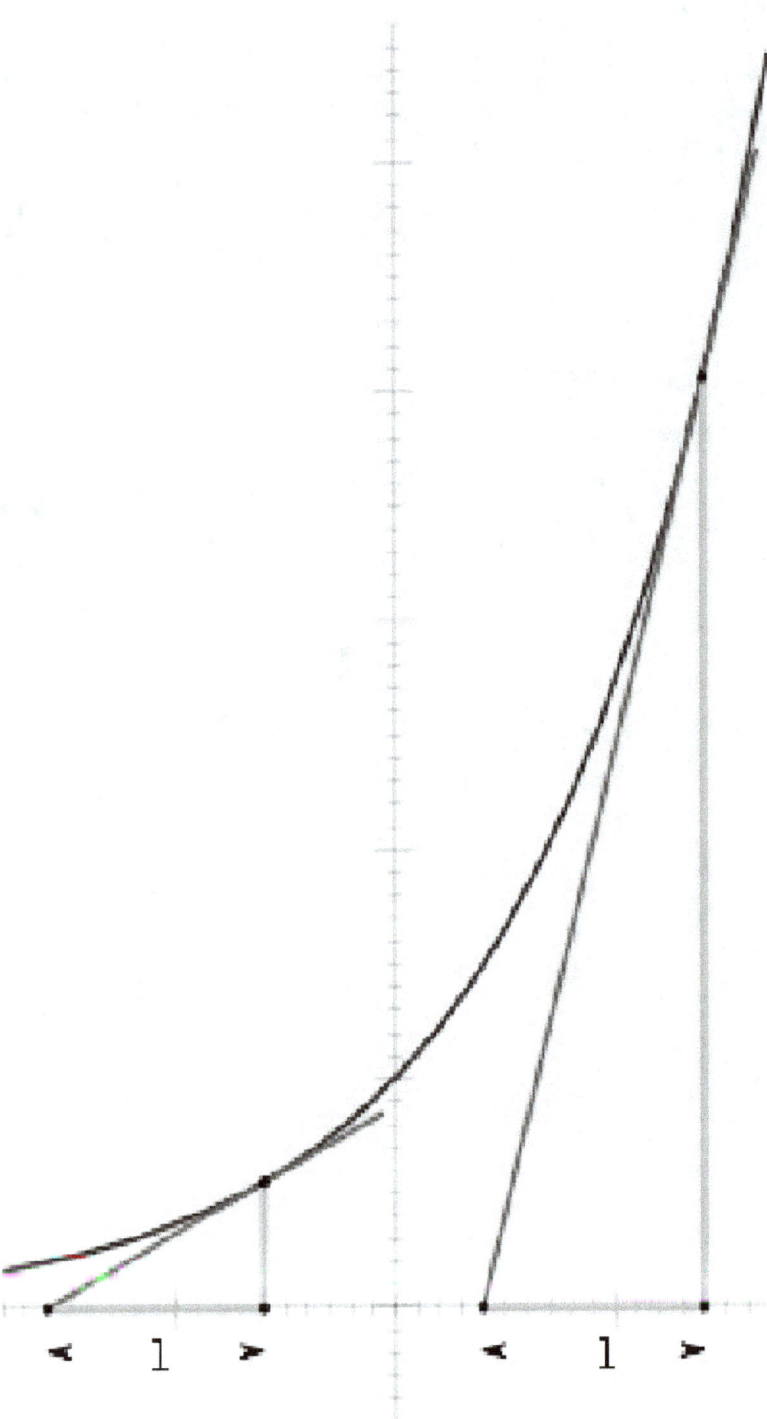

The derivative of the exponential function is equal to the value of the function. From any point P on the curve (blue), let a tangent line (red), and a vertical line (green) with height h be drawn, forming a right triangle with a base b on the x-axis. Since the slope of the red tangent line (the derivative) at P is equal to the ratio of the triangle's height to the triangle's base (rise over run), and the derivative is equal to the value of the function, h must be equal to the ratio of h to b. Therefore the base b must always be 1.

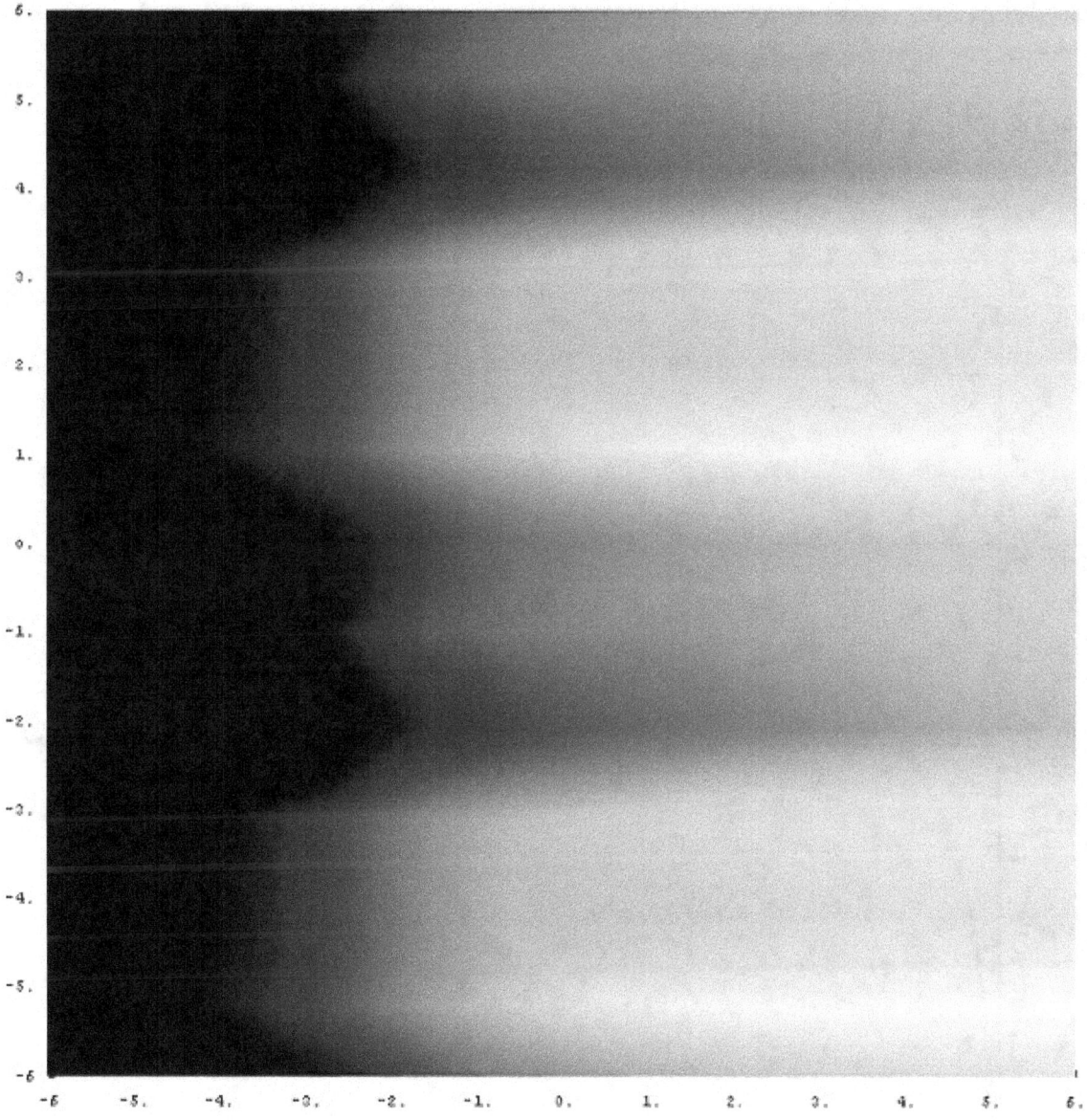

Exponential function on the complex plane. The transition from dark to light colors shows that the magnitude of the exponential function is increasing to the right. The periodic horizontal bands indicate that the exponential function is periodic in the imaginary part of its argument.

Chapter 16

Logarithm

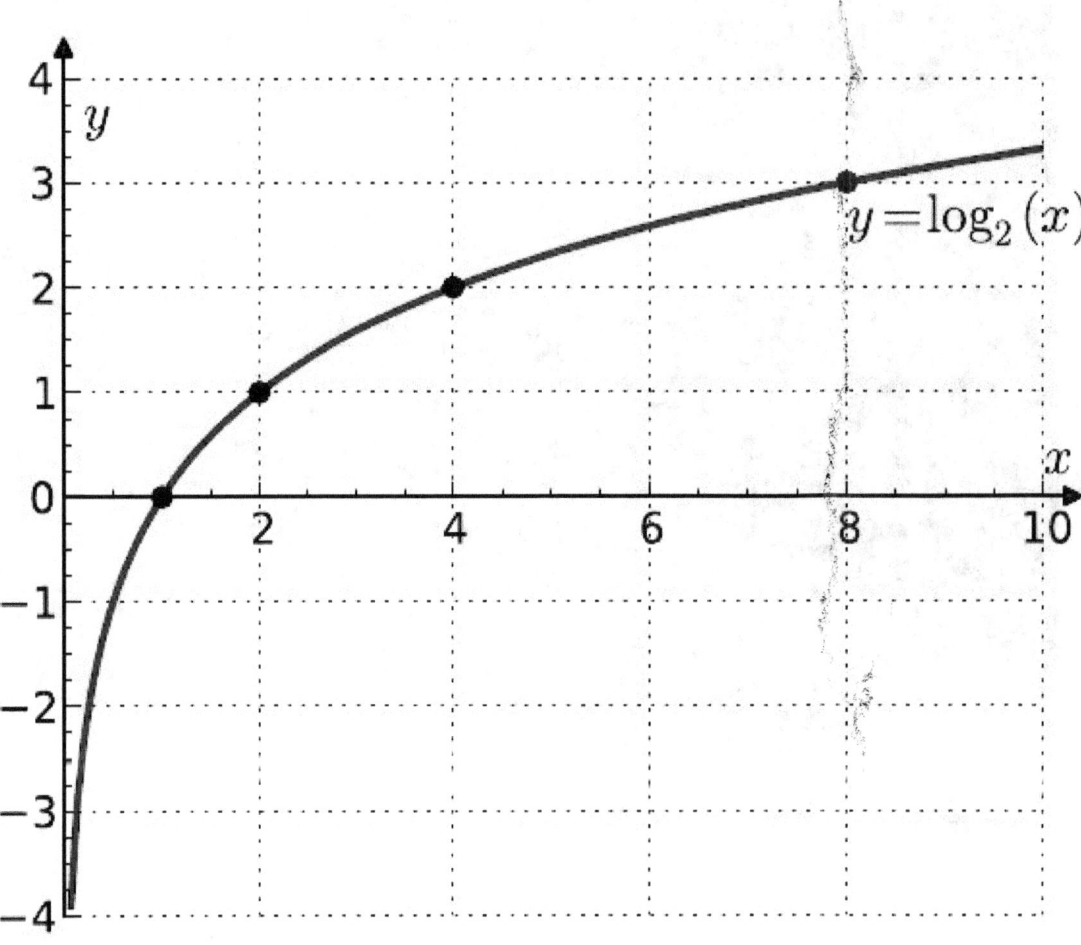

The graph of the logarithm to base 2 crosses the x axis (horizontal axis) at 1 and passes through the points with coordinates (2, 1), (4, 2), and (8, 3). For example, log₂(8) = 3, because 2³ = 8. The graph gets arbitrarily close to the y axis, but does not meet or intersect it.

In mathematics, the **logarithm** is the inverse operation to exponentiation. That means the logarithm of a number is the exponent to which another fixed value, the base, must be raised to produce that number. In simple cases the logarithm counts repeated multiplication. For example, the base 10 logarithm of 1000 is 3, as 10 to the power 3 is 1000 (1000 = 10 × 10 × 10 = 10³); the multiplication is repeated three times. More generally, exponentiation allows any positive real

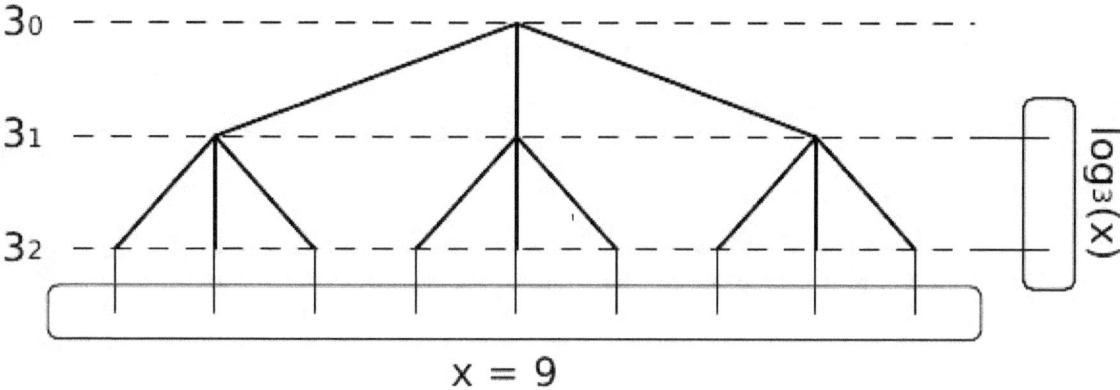

A full 3-ary tree can be used to visualize the exponents of 3 and how the logarithm function relates to them.

number to be raised to any real power, always producing a positive result, so the logarithm can be calculated for any two positive real numbers b and x where b is not equal to 1. The logarithm of x to *base b*, denoted $\log b(x)$, is the unique real number y such that

$$b^y = x.$$

For example, as $64 = 2^6$, we have

$$\log_2(64) = 6$$

The logarithm to base 10 (that is $b = 10$) is called the common logarithm and has many applications in science and engineering. The natural logarithm has the number e (≈ 2.718) as its base; its use is widespread in mathematics and physics, because of its simpler derivative. The binary logarithm uses base 2 (that is $b = 2$) and is commonly used in computer science.

Logarithms were introduced by John Napier in the early 17th century as a means to simplify calculations. They were rapidly adopted by navigators, scientists, engineers, and others to perform computations more easily, using slide rules and logarithm tables. Tedious multi-digit multiplication steps can be replaced by table look-ups and simpler addition because of the fact — important in its own right — that the logarithm of a product is the sum of the logarithms of the factors:

$$\log_b(xy) = \log_b(x) + \log_b(y),$$

provided that b, x and y are all positive and $b \neq 1$. The present-day notion of logarithms comes from Leonhard Euler, who connected them to the exponential function in the 18th century.

Logarithmic scales reduce wide-ranging quantities to tiny scopes. For example, the decibel is a unit quantifying signal power log-ratios and amplitude log-ratios (of which sound pressure is a common example). In chemistry, pH is a logarithmic measure for the acidity of an aqueous solution. Logarithms are commonplace in scientific formulae, and in measurements of the complexity of algorithms and of geometric objects called fractals. They describe musical intervals, appear in formulas counting prime numbers, inform some models in psychophysics, and can aid in forensic accounting.

In the same way as the logarithm reverses exponentiation, the complex logarithm is the inverse function of the exponential function applied to complex numbers. The discrete logarithm is another variant; it has uses in public-key cryptography.

16.1   Motivation and definition

The idea of logarithms is to reverse the operation of exponentiation, that is, raising a number to a power. For example, the third power (or cube) of 2 is 8, because 8 is the product of three factors of 2:

$2^3 = 2 \times 2 \times 2 = 8.$

It follows that the logarithm of 8 with respect to base 2 is 3, so $\log_2 8 = 3$.

16.1.1 Exponentiation

The third power of some number b is the product of three factors of b. More generally, raising b to the n-th power, where n is a natural number, is done by multiplying n factors of b. The n-th power of b is written b^n, so that

$$b^n = \underbrace{b \times b \times \cdots \times b}_{n\,\text{factors}}.$$

Exponentiation may be extended to b^y, where b is a positive number and the *exponent* y is any real number. For example, b^{-1} is the reciprocal of b, that is, $1/b$. (For further details, including the formula $b^{m+n} = b^m \cdot b^n$, see exponentiation or [1] for an elementary treatise.)

16.1.2 Definition

The *logarithm* of a positive real number x with respect to base b, a positive real number not equal to $1^{[\text{nb }1]}$, is the exponent by which b must be raised to yield x. In other words, the logarithm of x to base b is the solution y to the equation[2]

$$b^y = x.$$

The logarithm is denoted "$\log b(x)$" (pronounced as "the logarithm of x to base b" or "the base-b logarithm of x"). In the equation $y = \log b(x)$, the value y is the answer to the question "To what power must b be raised, in order to yield x?". This question can also be addressed (with a richer answer) for complex numbers, which is done in section "Complex logarithm", and this answer is much more extensively investigated in the page for the complex logarithm.

16.1.3 Examples

For example, $\log_2(16) = 4$, since $2^4 = 2 \times 2 \times 2 \times 2 = 16$. Logarithms can also be negative:

$$\log_2\left(\frac{1}{2}\right) = -1,$$

since

$$2^{-1} = \frac{1}{2^1} = \frac{1}{2}.$$

A third example: $\log_{10}(150)$ is approximately 2.176, which lies between 2 and 3, just as 150 lies between $10^2 = 100$ and $10^3 = 1000$. Finally, for any base b, $\log b(b) = 1$ and $\log b(1) = 0$, since $b^1 = b$ and $b^0 = 1$, respectively.

16.2 Logarithmic identities

Main article: List of logarithmic identities

Several important formulas, sometimes called *logarithmic identities* or *log laws*, relate logarithms to one another.[3]

16.2.1   Product, quotient, power and root

The logarithm of a product is the sum of the logarithms of the numbers being multiplied; the logarithm of the ratio of two numbers is the difference of the logarithms. The logarithm of the p-th power of a number is p times the logarithm of the number itself; the logarithm of a p-th root is the logarithm of the number divided by p. The following table lists these identities with examples. Each of the identities can be derived after substitution of the logarithm definitions $x = b^{\log_b(x)}$ or $y = b^{\log_b(y)}$ in the left hand sides.

16.2.2   Change of base

The logarithm $\log_b(x)$ can be computed from the logarithms of x and b with respect to an arbitrary base k using the following formula:

$$\log_b(x) = \frac{\log_k(x)}{\log_k(b)}.$$

Typical scientific calculators calculate the logarithms to bases 10 and e.[4] Logarithms with respect to any base b can be determined using either of these two logarithms by the previous formula:

$$\log_b(x) = \frac{\log_{10}(x)}{\log_{10}(b)} = \frac{\log_e(x)}{\log_e(b)}.$$

Given a number x and its logarithm $\log_b(x)$ to an unknown base b, the base is given by:

$$b = x^{\frac{1}{\log_b(x)}}.$$

16.3   Particular bases

Among all choices for the base, three are particularly common. These are $b = 10$, $b = e$ (the irrational mathematical constant ≈ 2.71828), and $b = 2$. In mathematical analysis, the logarithm to base e is widespread because of its particular analytical properties explained below. On the other hand, base-10 logarithms are easy to use for manual calculations in the decimal number system:[5]

$$\log_{10}(10x) = \log_{10}(10) + \log_{10}(x) = 1 + \log_{10}(x).$$

Thus, $\log_{10}(x)$ is related to the number of decimal digits of a positive integer x: the number of digits is the smallest integer strictly bigger than $\log_{10}(x)$.[6] For example, $\log_{10}(1430)$ is approximately 3.15. The next integer is 4, which is the number of digits of 1430. Both the natural logarithm and the logarithm to base two are used in information theory, corresponding to the use of nats or bits as the fundamental units of information, respectively.[7] Binary logarithms are also used in computer science, where the binary system is ubiquitous, in music theory, where a pitch ratio of two (the octave) is ubiquitous and the cent is the binary logarithm (scaled by 1200) of the ratio between two adjacent equally-tempered pitches, and in photography to measure exposure values.[8]

The following table lists common notations for logarithms to these bases and the fields where they are used. Many disciplines write $\log(x)$ instead of $\log_b(x)$, when the intended base can be determined from the context. The notation $^b\log(x)$ also occurs.[9] The "ISO notation" column lists designations suggested by the International Organization for Standardization (ISO 31-11).[10]

16.4 History

16.4.1 Predecessors

The Babylonians sometime in 2000–1600 BC may have invented the quarter square multiplication algorithm to multiply two numbers using only addition, subtraction and a table of quarter squares.[17][18] However, it could not be used for division without an additional table of reciprocals (or the knowledge of a sufficiently simple algorithm to generate reciprocals). Large tables of quarter squares were used to simplify the accurate multiplication of large numbers from 1817 onwards until this was superseded by the use of computers.

The Indian mathematician Virasena worked with the concept of ardhaccheda: the number of times a number of the form 2n could be halved. For exact powers of 2, this is the logarithm to that base, which is a whole number; for other numbers, it is undefined. He described relations such as the product formula and also introduced integer logarithms in base 3 (trakacheda) and base 4 (caturthacheda).[19]

Michael Stifel published *Arithmetica integra* in Nuremberg in 1544, which contains a table[20] of integers and powers of 2 that has been considered an early version of a logarithmic table.[21][22]

In the 16th and early 17th centuries an algorithm called prosthaphaeresis was used to approximate multiplication and division. This used the trigonometric identity

$$\cos\alpha\,\cos\beta = \frac{1}{2}[\cos(\alpha+\beta)+\cos(\alpha-\beta)]$$

or similar to convert the multiplications to additions and table lookups. However, logarithms are more straightforward and require less work. It can be shown using Euler's Formula that the two techniques are related.

16.4.2 From Napier to Euler

The method of logarithms was publicly propounded by John Napier in 1614, in a book titled *Mirifici Logarithmorum Canonis Descriptio* (*Description of the Wonderful Rule of Logarithms*).[23][24] Joost Bürgi constructed a table of powers with a basis very close to 1, and this table provides a fine correspondence between the integers 1-10 (or 10-100, etc.) and exponents that can be added. This table was printed (but perhaps not published) in 1620. However, Bürgi did not define an abstract continuous function as Napier did, and he also did not work out the accuracy of interpolations, which was also tackled by Napier.[25][26]

Johannes Kepler, who used logarithm tables extensively to compile his *Ephemeris* and therefore dedicated it to Napier,[27] remarked:

> ... the accent in calculation led Justus Byrgius [Joost Bürgi] on the way to these very logarithms many years before Napier's system appeared; but ... instead of rearing up his child for the public benefit he deserted it in the birth.
> — Johannes Kepler[28], Rudolphine Tables (1627)

By repeated subtractions Napier calculated $(1-10^{-7})^L$ for L ranging from 1 to 100. The result for $L=100$ is approximately $0.99999 = 1-10^{-5}$. Napier then calculated the products of these numbers with $10^7(1-10^{-5})^L$ for L from 1 to 50, and did similarly with $0.9998 \approx (1-10^{-5})^{20}$ and $0.9 \approx 0.995^{20}$. These computations, which occupied 20 years, allowed him to give, for any number N from 5 to 10 million, the number L that solves the equation

$$N = 10^7(1-10^{-7})^L.$$

Napier first called L an "artificial number", but later introduced the word *"logarithm"* to mean a number that indicates a ratio: λόγος (logos) meaning proportion, and ἀριθμός (arithmos) meaning number. In modern notation, the relation to natural logarithms is: [29]

John Napier (1550–1617), the inventor of logarithms

$$L = \log_{(1-10^{-7})}\left(\frac{N}{10^7}\right) \approx 10^7 \log_{\frac{1}{7}}\left(\frac{N}{10^7}\right) = -10^7 \log_e\left(\frac{N}{10^7}\right).$$

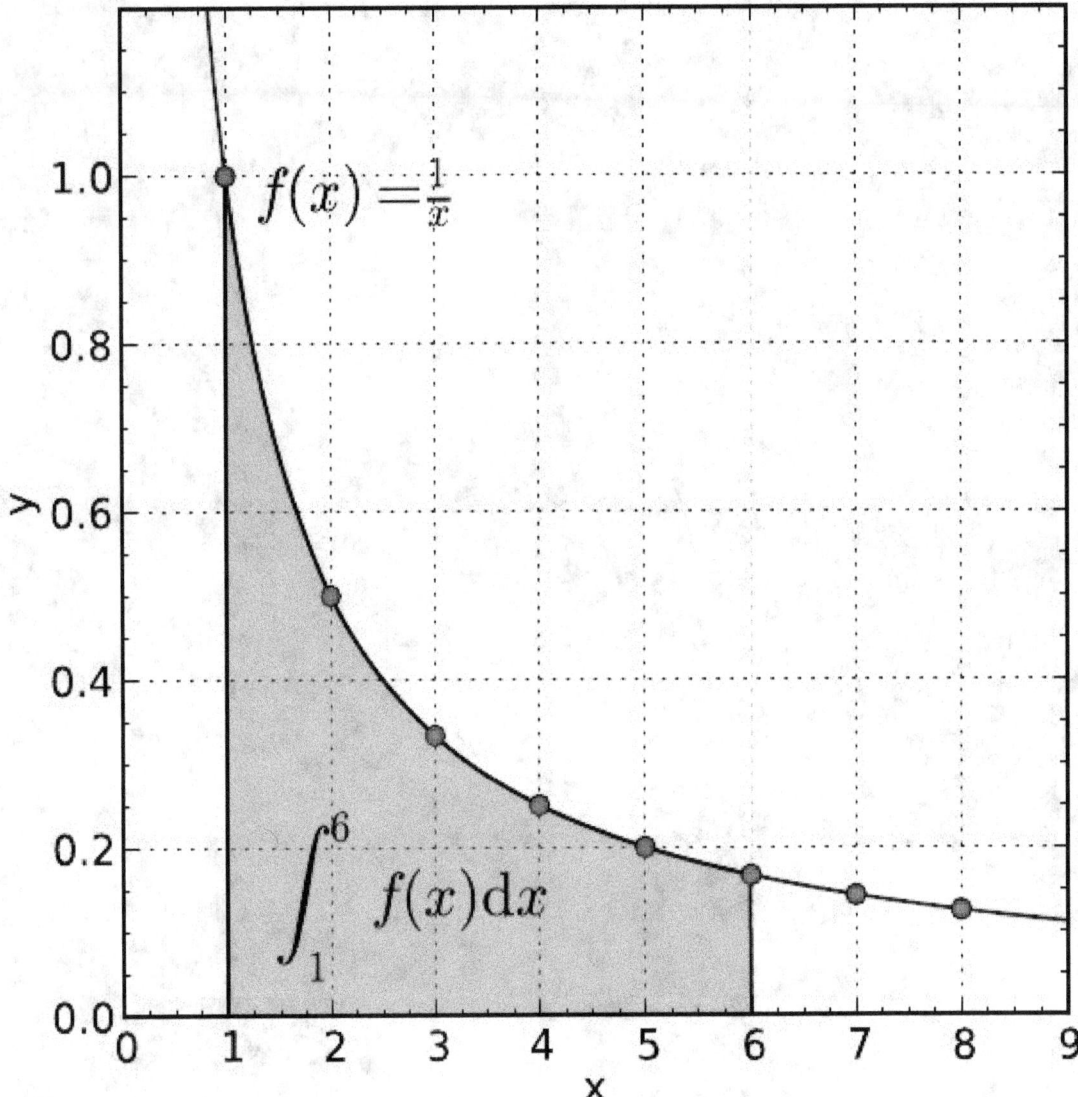

The hyperbola y = 1/x *(red curve) and the area from* x = 1 *to 6 (shaded in orange). This area equals the natural logarithm of 6.*

where the very close approximation corresponds to the observation that

$$(1 - 10^{-7})^{10^7} \approx \frac{1}{e}.$$

The invention was quickly and widely met with acclaim. The works of Bonaventura Cavalieri (Italy), Edmund Wingate (France), Xue Fengzuo (China), and Johannes Kepler's *Chilias logarithmorum* (Germany) helped spread the concept further.[30]

In 1649, Alphonse Antonio de Sarasa, a former student of Grégoire de Saint-Vincent,[31] related logarithms to the quadrature of the hyperbola, by pointing out that the area $A(t)$ under the hyperbola from $x = 1$ to $x = t$ satisfies[32]

$$A(tu) = A(t) + A(u).$$

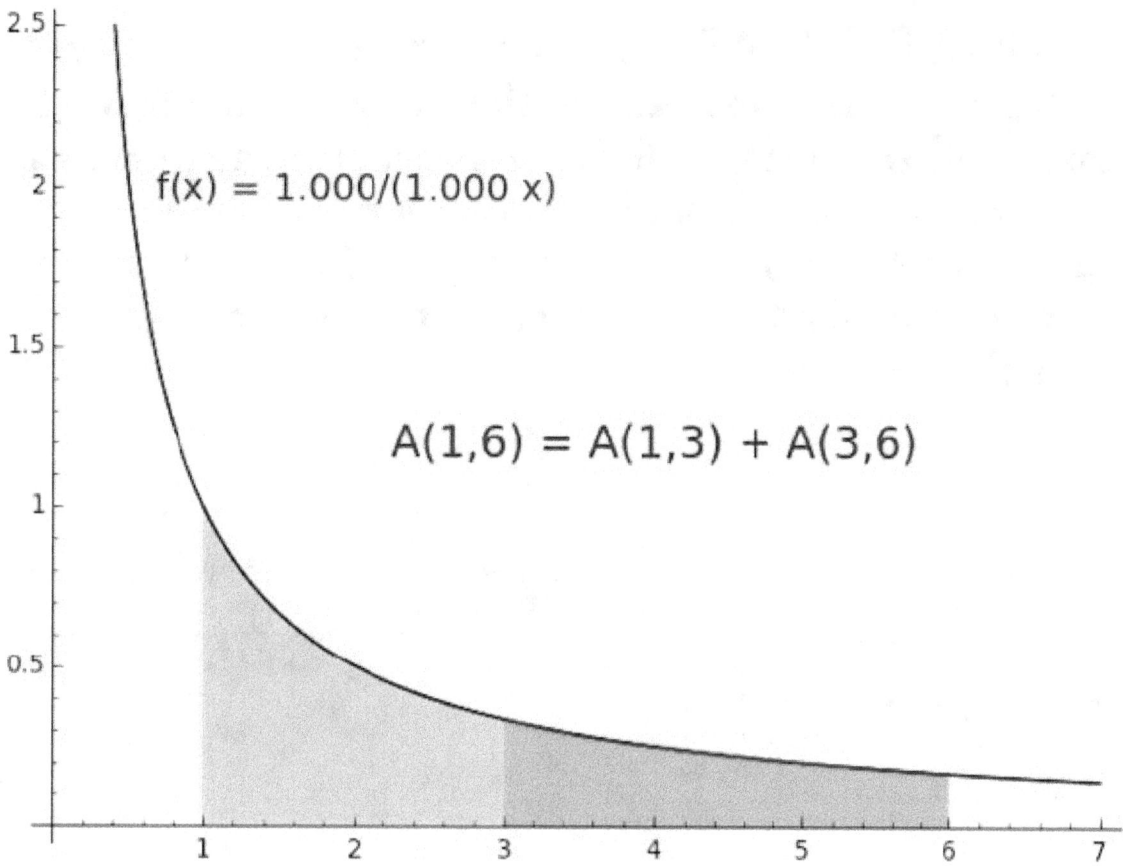

The area under the hyperbola satisfies the logarithm rule. Here A(a,b) denotes the area under the hyperbola between a and b.

The natural logarithm was first described by Nicholas Mercator in his work *Logarithmotechnia* published in 1668,[33] although the mathematics teacher John Speidell had already in 1619 compiled a table of what were effectively natural logarithms, based on Napier's work.[34] Around 1730, Leonhard Euler defined the exponential function and the natural logarithm by

$$e^x = \lim_{n \to \infty} (1 + x/n)^n.$$

$$\ln(x) = \lim_{n \to \infty} n(x^{1/n} - 1).$$

Euler also showed that the two functions are inverse to one another.[35][36][37]

16.4.3   Logarithm tables, slide rules, and historical applications

By simplifying difficult calculations, logarithms contributed to the advance of science, and especially of astronomy. They were critical to advances in surveying, celestial navigation, and other domains. Pierre-Simon Laplace called logarithms

> "...[a]n admirable artifice which, by reducing to a few days the labour of many months, doubles the life of the astronomer, and spares him the errors and disgust inseparable from long calculations."[38]

A key tool that enabled the practical use of logarithms before calculators and computers was the *table of logarithms*.[39] The first such table was compiled by Henry Briggs in 1617, immediately after Napier's invention. Subsequently, tables

LOGARITHMS, (from λoγ• *ratio,* and αριθμ•• *number*), the indices of the ratios of numbers to one another ; being a feries of numbers in arithmetical progreffion, correfponding to others in geometrical progreffion ; by means of which, arithmetical calculations can be made with much more eafe and expedition than otherwife.

The 1797 Encyclopædia Britannica *explanation of logarithms*

with increasing scope were written. These tables listed the values of $\log_b(x)$ and b^x for any number x in a certain range, at a certain precision, for a certain base b (usually $b = 10$). For example, Briggs' first table contained the common logarithms of all integers in the range 1–1000, with a precision of 14 digits. As the function $f(x) = b^x$ is the inverse function of $\log_b(x)$, it has been called the **antilogarithm**.[40] The product and quotient of two positive numbers c and d were routinely calculated as the sum and difference of their logarithms. The product cd or quotient c/d came from looking up the antilogarithm of the sum or difference, also via the same table:

$$cd = b^{\log_b(c)} b^{\log_b(d)} = b^{\log_b(c) + \log_b(d)}$$

and

$$\frac{c}{d} = cd^{-1} = b^{\log_b(c) - \log_b(d)}.$$

For manual calculations that demand any appreciable precision, performing the lookups of the two logarithms, calculating their sum or difference, and looking up the antilogarithm is much faster than performing the multiplication by earlier methods such as prosthaphaeresis, which relies on trigonometric identities. Calculations of powers and roots are reduced to multiplications or divisions and look-ups by

$$c^d = \left(b^{\log_b(c)}\right)^d = b^{d \log_b(c)}$$

and

$$\sqrt[d]{c} = c^{\frac{1}{d}} = b^{\frac{1}{d} \log_b(c)}.$$

Many logarithm tables give logarithms by separately providing the characteristic and mantissa of x, that is to say, the integer part and the fractional part of $\log_{10}(x)$.[41] The characteristic of $10 \cdot x$ is one plus the characteristic of x, and their significands are the same. This extends the scope of logarithm tables: given a table listing $\log_{10}(x)$ for all integers x ranging from 1 to 1000, the logarithm of 3542 is approximated by

$$\log_{10}(3542) = \log_{10}(10 \cdot 354.2) = 1 + \log_{10}(354.2) \approx 1 + \log_{10}(354). \text{ Greater accuracy can be obtained by interpolation.}$$

Another critical application was the slide rule, a pair of logarithmically divided scales used for calculation, as illustrated here:

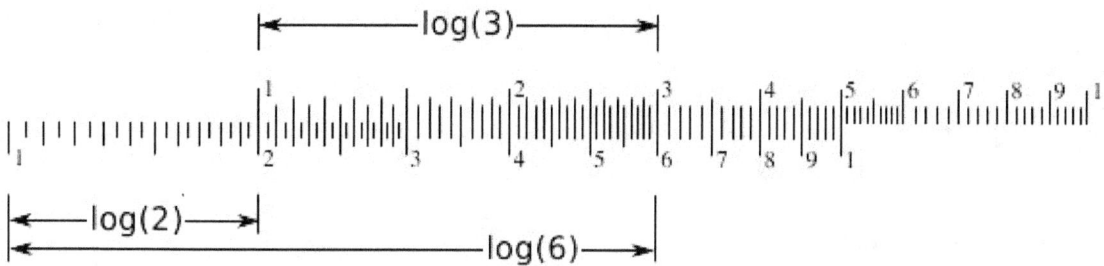

Schematic depiction of a slide rule. Starting from 2 on the lower scale, add the distance to 3 on the upper scale to reach the product 6. The slide rule works because it is marked such that the distance from 1 to x is proportional to the logarithm of x.

The non-sliding logarithmic scale, Gunter's rule, was invented shortly after Napier's invention. William Oughtred enhanced it to create the slide rule—a pair of logarithmic scales movable with respect to each other. Numbers are placed on sliding scales at distances proportional to the differences between their logarithms. Sliding the upper scale appropriately amounts to mechanically adding logarithms. For example, adding the distance from 1 to 2 on the lower scale to the distance from 1 to 3 on the upper scale yields a product of 6, which is read off at the lower part. The slide rule was an essential calculating tool for engineers and scientists until the 1970s, because it allows, at the expense of precision, much faster computation than techniques based on tables.[35]

16.5 Analytic properties

A deeper study of logarithms requires the concept of a *function*. A function is a rule that, given one number, produces another number.[42] An example is the function producing the *x*-th power of *b* from any real number *x*, where the base *b* is a fixed number. This function is written

$$f(x) = b^x.$$

16.5.1 Logarithmic function

To justify the definition of logarithms, it is necessary to show that the equation

$$b^x = y$$

has a solution *x* and that this solution is unique, provided that *y* is positive and that *b* is positive and unequal to 1. A proof of that fact requires the intermediate value theorem from elementary calculus.[43] This theorem states that a continuous function that produces two values *m* and *n* also produces any value that lies between *m* and *n*. A function is *continuous* if it does not "jump", that is, if its graph can be drawn without lifting the pen.

This property can be shown to hold for the function $f(x) = b^x$. Because f takes arbitrarily large and arbitrarily small positive values, any number $y > 0$ lies between $f(x_0)$ and $f(x_1)$ for suitable x_0 and x_1. Hence, the intermediate value theorem ensures that the equation $f(x) = y$ has a solution. Moreover, there is only one solution to this equation, because the function f is strictly increasing (for $b > 1$), or strictly decreasing (for $0 < b < 1$).[44]

The unique solution *x* is the logarithm of *y* to base *b*, $\log_b(y)$. The function that assigns to *y* its logarithm is called *logarithm function* or *logarithmic function* (or just *logarithm*).

The function $\log_b(x)$ is essentially characterized by the above product formula

$$\log_b(xy) = \log_b(x) + \log_b(y).$$

More precisely, the logarithm to any base $b > 1$ is the only increasing function f from the positive reals to the reals satisfying $f(b) = 1$ and [45]

$$f(xy) = f(x) + f(y).$$

16.5.2 Inverse function

The formula for the logarithm of a power says in particular that for any number x,

$$\log_b (b^x) = x \log_b (b) = x.$$

In prose, taking the x-th power of b and then the base-b logarithm gives back x. Conversely, given a positive number y, the formula

$$b^{\log_b (y)} = y$$

says that first taking the logarithm and then exponentiating gives back y. Thus, the two possible ways of combining (or composing) logarithms and exponentiation give back the original number. Therefore, the logarithm to base b is the *inverse function* of $f(x) = b^x$.[46]

Inverse functions are closely related to the original functions. Their graphs correspond to each other upon exchanging the x- and the y-coordinates (or upon reflection at the diagonal line $x = y$), as shown at the right: a point $(t, u = b^t)$ on the graph of f yields a point $(u, t = \log_b u)$ on the graph of the logarithm and vice versa. As a consequence, $\log b(x)$ diverges to infinity (gets bigger than any given number) if x grows to infinity, provided that b is greater than one. In that case, $\log b(x)$ is an increasing function. For $b < 1$, $\log b(x)$ tends to minus infinity instead. When x approaches zero, $\log b(x)$ goes to minus infinity for $b > 1$ (plus infinity for $b < 1$, respectively).

16.5.3 Derivative and antiderivative

Analytic properties of functions pass to their inverses.[43] Thus, as $f(x) = b^x$ is a continuous and differentiable function, so is $\log b(y)$. Roughly, a continuous function is differentiable if its graph has no sharp "corners". Moreover, as the derivative of $f(x)$ evaluates to $\ln(b)b^x$ by the properties of the exponential function, the chain rule implies that the derivative of $\log b(x)$ is given by[44][47]

$$\frac{d}{dx} \log_b (x) = \frac{1}{x \ln(b)}.$$

That is, the slope of the tangent touching the graph of the base-b logarithm at the point $(x, \log b(x))$ equals $1/(x \ln(b))$.

The derivative of $\ln(x)$ is $1/x$; this implies that $\ln(x)$ is the unique antiderivative of $1/x$ that has the value 0 for $x = 1$. This is this very simple formula that motivated to qualify as "natural" the natural logarithm; this is also one of the main reasons of the importance of the constant e.

The derivative with a generalised functional argument $f(x)$ is

$$\frac{d}{dx} \ln(f(x)) = \frac{f'(x)}{f(x)}.$$

The quotient at the right hand side is called the logarithmic derivative of f. Computing $f'(x)$ by means of the derivative of $\ln(f(x))$ is known as logarithmic differentiation.[48] The antiderivative of the natural logarithm $\ln(x)$ is:[49]

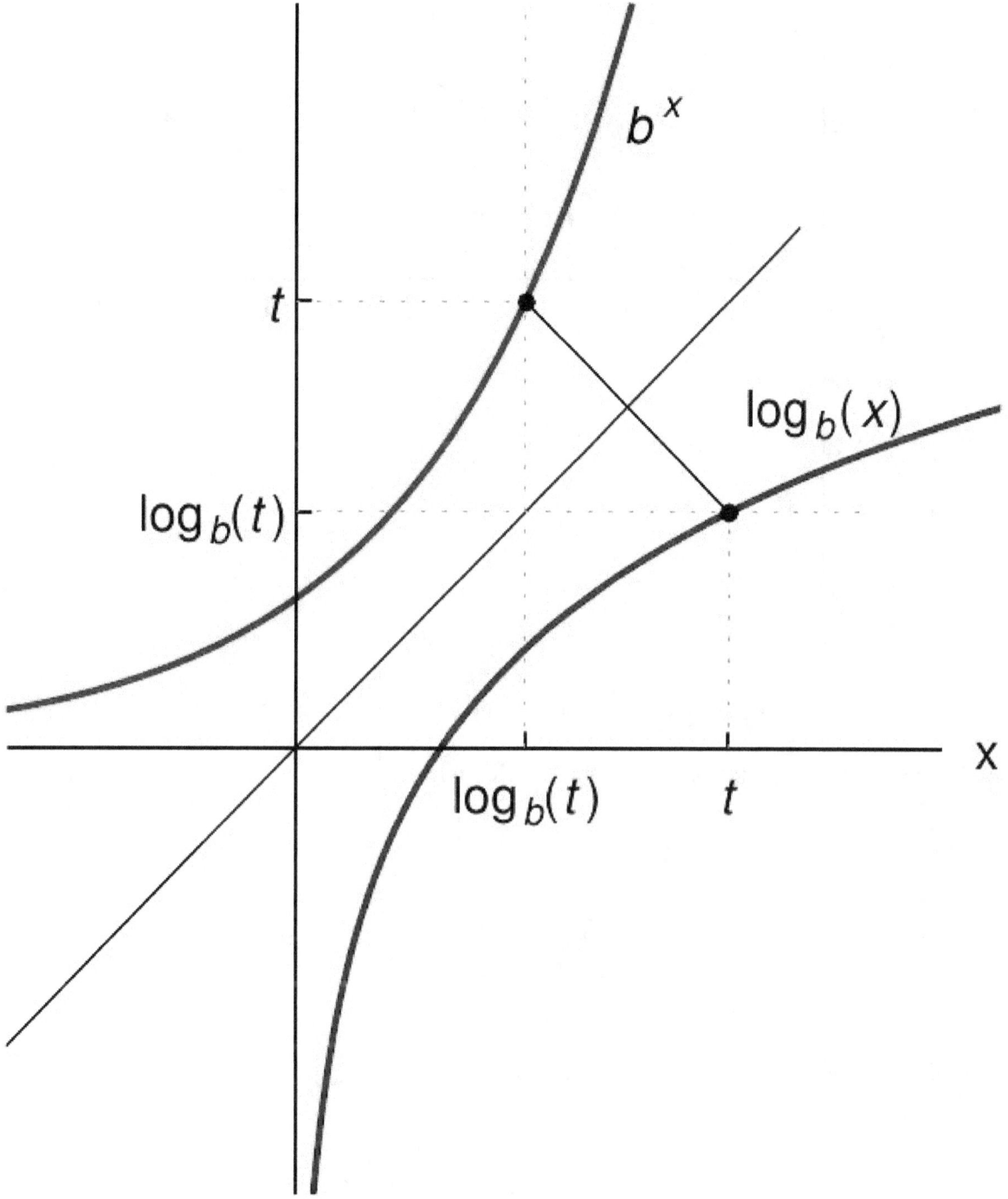

The graph of the logarithm function $\log_b(x)$ (blue) is obtained by reflecting the graph of the function b^x (red) at the diagonal line (x = y).

$$\int \ln(x)\,dx = x\ln(x) - x + C.$$

Related formulas, such as antiderivatives of logarithms to other bases can be derived from this equation using the change of bases.[50]

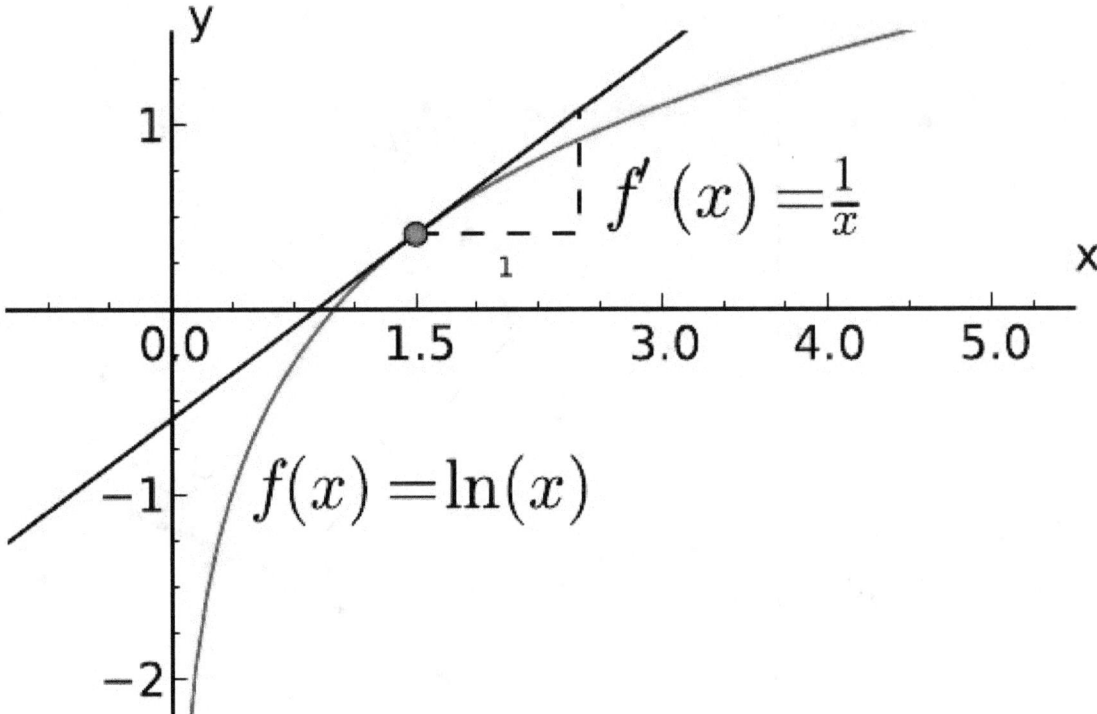

The graph of the natural logarithm (green) and its tangent at x = 1.5 (black)

16.5.4   Integral representation of the natural logarithm

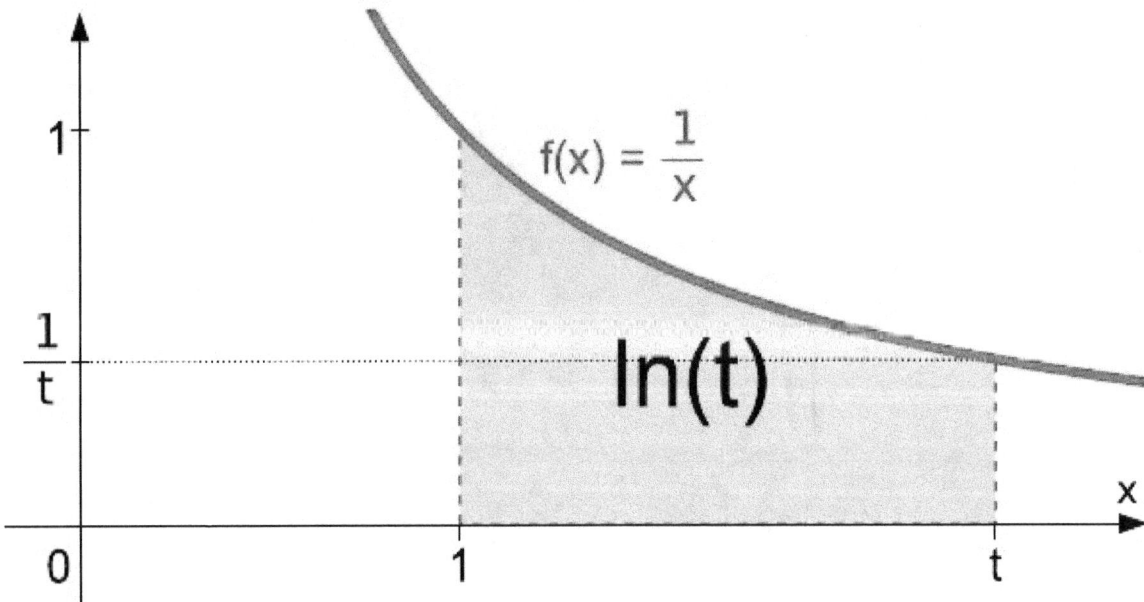

The natural logarithm of t is the shaded area underneath the graph of the function f(x) = 1/x (reciprocal of x).

The natural logarithm of t equals the integral of $1/x\ dx$ from 1 to t:

$$\ln(t) = \int_1^t \tfrac{1}{x}\,dx.$$

In other words, ln(t) equals the area between the x axis and the graph of the function $1/x$, ranging from $x = 1$ to $x = t$ (figure at the right). This is a consequence of the fundamental theorem of calculus and the fact that derivative of ln(x) is $1/x$. The right hand side of this equation can serve as a definition of the natural logarithm. Product and power logarithm formulas can be derived from this definition.[51] For example, the product formula ln(tu) = ln(t) + ln(u) is deduced as:

$$\ln(tu) = \int_1^{tu} \frac{1}{x}\,dx \stackrel{(1)}{=} \int_1^t \frac{1}{x}\,dx + \int_t^{tu} \frac{1}{x}\,dx \stackrel{(2)}{=} \ln(t) + \int_1^u \frac{1}{w}\,dw = \ln(t) + \ln(u).$$

The equality (1) splits the integral into two parts, while the equality (2) is a change of variable ($w = x/t$). In the illustration below, the splitting corresponds to dividing the area into the yellow and blue parts. Rescaling the left hand blue area vertically by the factor t and shrinking it by the same factor horizontally does not change its size. Moving it appropriately, the area fits the graph of the function $f(x) = 1/x$ again. Therefore, the left hand blue area, which is the integral of $f(x)$ from t to tu is the same as the integral from 1 to u. This justifies the equality (2) with a more geometric proof.

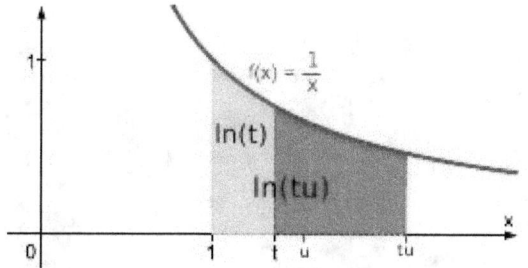 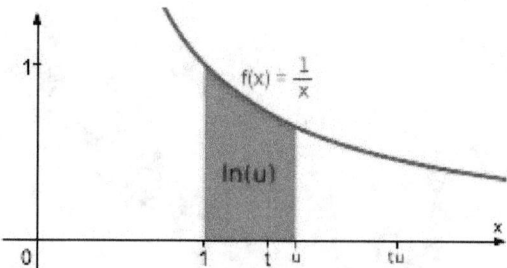

A visual proof of the product formula of the natural logarithm

The power formula ln(t^r) = r ln(t) may be derived in a similar way:

$$\ln(t^r) = \int_1^{t^r} \frac{1}{x}\,dx = \int_1^t \frac{1}{w^r}\left(rw^{r-1}\,dw\right) = r\int_1^t \frac{1}{w}\,dw = r\ln(t).$$

The second equality uses a change of variables (integration by substitution), $w = x^{1/r}$.

The sum over the reciprocals of natural numbers,

$$1 + \frac{1}{2} + \frac{1}{3} + \cdots + \frac{1}{n} = \sum_{k=1}^{n} \frac{1}{k}.$$

is called the harmonic series. It is closely tied to the natural logarithm: as n tends to infinity, the difference,

$$\sum_{k=1}^{n} \frac{1}{k} - \ln(n).$$

converges (i.e., gets arbitrarily close) to a number known as the Euler–Mascheroni constant. This relation aids in analyzing the performance of algorithms such as quicksort.[52]

There is also another integral representation of the logarithm that is useful in some situations.

$$\ln(x) = -\lim_{\epsilon \to 0} \int_{\epsilon}^{\infty} \frac{dt}{t}\left(e^{-xt} - e^{-t}\right)$$

This can be verified by showing that it has the same value at $x = 1$, and the same derivative.

16.5.5   Transcendence of the logarithm

Real numbers that are not algebraic are called transcendental;[53] for example, π and e are such numbers, but $\sqrt{2 - \sqrt{3}}$ is not. Almost all real numbers are transcendental. The logarithm is an example of a transcendental function. The Gelfond–Schneider theorem asserts that logarithms usually take transcendental, i.e., "difficult" values.[54]

16.6   Calculation

The logarithm keys (log for base-10 and ln for base-e) on a typical scientific calculator

Logarithms are easy to compute in some cases, such as $\log_{10}(1{,}000) = 3$. In general, logarithms can be calculated using power series or the arithmetic–geometric mean, or be retrieved from a precalculated logarithm table that provides a fixed precision.[55][56] Newton's method, an iterative method to solve equations approximately, can also be used to calculate the logarithm, because its inverse function, the exponential function, can be computed efficiently.[57] Using look-up tables, CORDIC-like methods can be used to compute logarithms if the only available operations are addition and bit shifts.[58][59] Moreover, the binary logarithm algorithm calculates lb(x) recursively based on repeated squarings of x, taking advantage of the relation

$$\log_2(x^2) = 2\log_2(x).$$

16.6.1   Power series

Taylor series

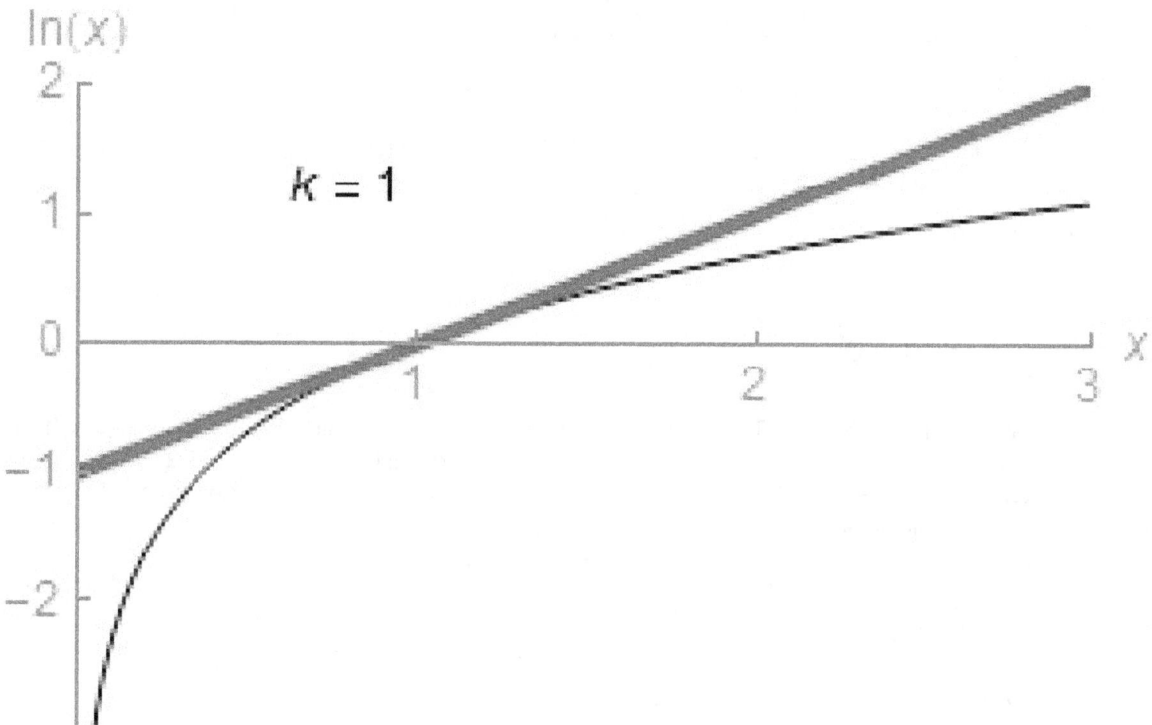

The Taylor series of ln(z) centered at z = 1. The animation shows the first 10 approximations along with the 99th and 100th. The approximations do not converge beyond a distance of 1 from the center.

For any real number z that satisfies $0 < z < 2$, the following formula holds:[nb 4][60]

$$\ln(z) = \frac{(z-1)^1}{1} - \frac{(z-1)^2}{2} + \frac{(z-1)^3}{3} - \frac{(z-1)^4}{4} + \cdots$$

This is a shorthand for saying that $\ln(z)$ can be approximated to a more and more accurate value by the following expressions:

$$(z-1)$$
$$(z-1) \quad - \quad \frac{(z-1)^2}{2}$$
$$(z-1) \quad - \quad \frac{(z-1)^2}{2} \quad + \quad \frac{(z-1)^3}{3}$$
$$\vdots$$

For example, with $z = 1.5$ the third approximation yields 0.4167, which is about 0.011 greater than $\ln(1.5) = 0.405465$. This series approximates $\ln(z)$ with arbitrary precision, provided the number of summands is large enough. In elementary calculus, $\ln(z)$ is therefore the *limit* of this series. It is the Taylor series of the natural logarithm at $z = 1$. The Taylor series of $\ln z$ provides a particularly useful approximation to $\ln(1+z)$ when z is small, $|z| < 1$, since then

$$\ln(1 + z) = z - \frac{z^2}{2} + \frac{z^3}{3} \cdots \approx z.$$

For example, with $z = 0.1$ the first-order approximation gives $\ln(1.1) \approx 0.1$, which is less than 5% off the correct value 0.0953.

More efficient series

Another series is based on the area hyperbolic tangent function:

$$\ln(z) = 2 \cdot \text{artanh}\, \frac{z-1}{z+1} = 2\left(\frac{z-1}{z+1} + \frac{1}{3}\left(\frac{z-1}{z+1}\right)^3 + \frac{1}{5}\left(\frac{z-1}{z+1}\right)^5 + \cdots\right),$$

for any real number $z > 0$.[nb 5][60] Using the Sigma notation, this is also written as

$$\ln(z) = 2\sum_{n=0}^{\infty} \frac{1}{2n+1}\left(\frac{z-1}{z+1}\right)^{2n+1}.$$

This series can be derived from the above Taylor series. It converges more quickly than the Taylor series, especially if z is close to 1. For example, for $z = 1.5$, the first three terms of the second series approximate $\ln(1.5)$ with an error of about 3×10^{-6}. The quick convergence for z close to 1 can be taken advantage of in the following way: given a low-accuracy approximation $y \approx \ln(z)$ and putting

$$A = \frac{z}{\exp(y)},$$

the logarithm of z is:

$$\ln(z) = y + \ln(A).$$

The better the initial approximation y is, the closer A is to 1, so its logarithm can be calculated efficiently. A can be calculated using the exponential series, which converges quickly provided y is not too large. Calculating the logarithm of larger z can be reduced to smaller values of z by writing $z = a \cdot 10^b$, so that $\ln(z) = \ln(a) + b \cdot \ln(10)$.

A closely related method can be used to compute the logarithm of integers. From the above series, it follows that:

$$\ln(n+1) = \ln(n) + 2\sum_{k=0}^{\infty} \frac{1}{2k+1}\left(\frac{1}{2n+1}\right)^{2k+1}.$$

If the logarithm of a large integer n is known, then this series yields a fast converging series for $\log(n+1)$.

16.6.2 Arithmetic–geometric mean approximation

The arithmetic–geometric mean yields high precision approximations of the natural logarithm. $\ln(x)$ is approximated to a precision of 2^{-p} (or p precise bits) by the following formula (due to Carl Friedrich Gauss):[61][62]

$$\ln(x) \approx \frac{\pi}{2M(1, 2^{2-m}/x)} - m\ln(2).$$

Here $M(x,y)$ denotes the arithmetic–geometric mean of x and y. It is obtained by repeatedly calculating the average $(x+y)/2$ (arithmetic mean) and sqrt(x*y) (geometric mean) of x and y then let those two numbers become the next x and y. The two numbers quickly converge to a common limit which is the value of $M(x,y)$. m is chosen such that

$x\,2^m > 2^{p/2}.$

to insure the required precision. A larger m makes the $M(x,y)$ calculation take more steps (the initial x and y are farther apart so it takes more steps to converge) but gives more precision. The constants π and $\ln(2)$ can be calculated with quickly converging series.

16.7 Applications

A nautilus displaying a logarithmic spiral

Logarithms have many applications inside and outside mathematics. Some of these occurrences are related to the notion of scale invariance. For example, each chamber of the shell of a nautilus is an approximate copy of the next one, scaled by a constant factor. This gives rise to a logarithmic spiral.[63] Benford's law on the distribution of leading digits can also be explained by scale invariance.[64] Logarithms are also linked to self-similarity. For example, logarithms appear in the analysis of algorithms that solve a problem by dividing it into two similar smaller problems and patching their solutions.[65] The dimensions of self-similar geometric shapes, that is, shapes whose parts resemble the overall picture are also based on logarithms. Logarithmic scales are useful for quantifying the relative change of a value as opposed to its absolute difference. Moreover, because the logarithmic function $\log(x)$ grows very slowly for large x, logarithmic scales are used to compress large-scale scientific data. Logarithms also occur in numerous scientific formulas, such as the Tsiolkovsky rocket equation, the Fenske equation, or the Nernst equation.

16.7.1 Logarithmic scale

Main article: Logarithmic scale

Scientific quantities are often expressed as logarithms of other quantities, using a *logarithmic scale*. For example, the

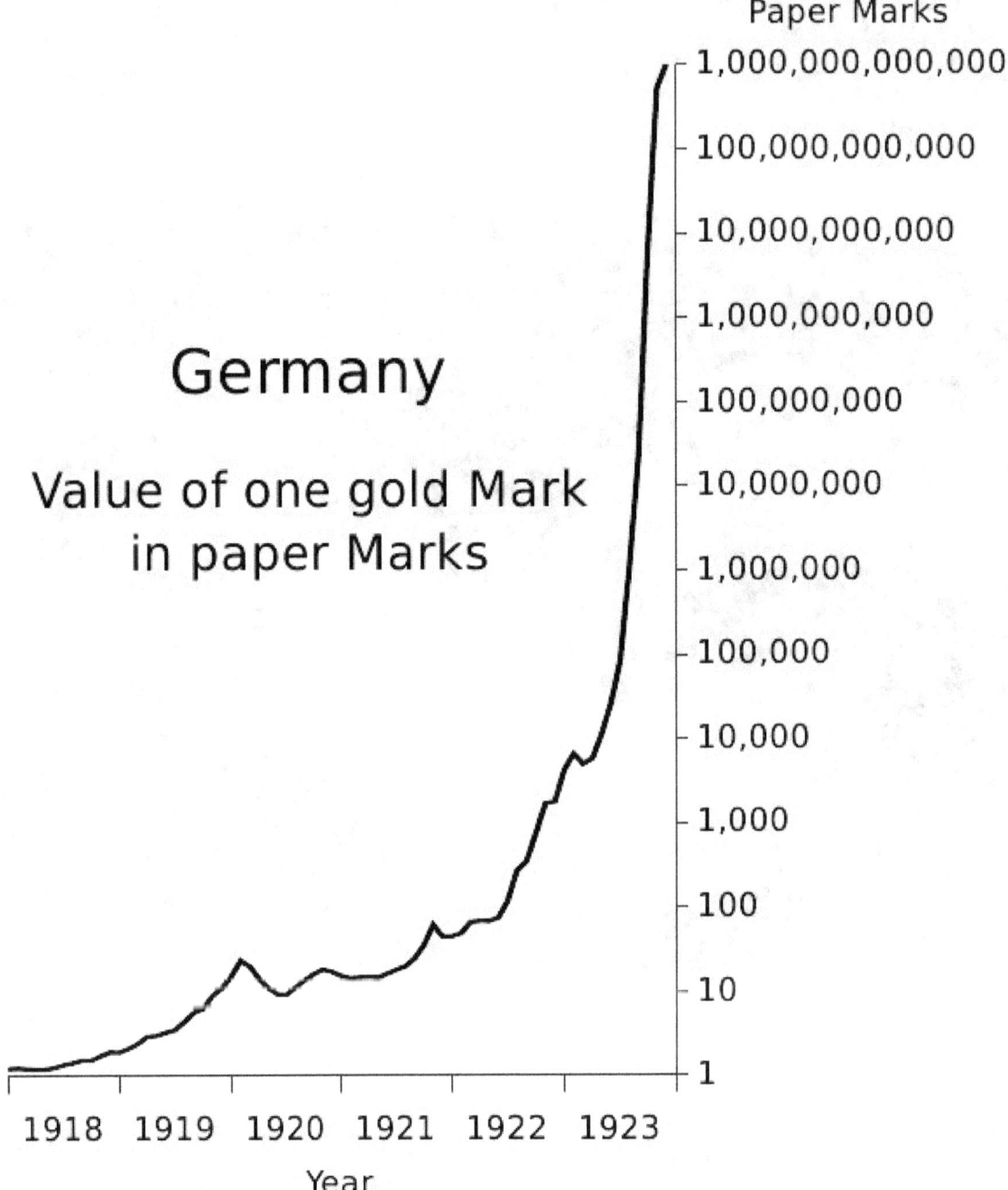

A logarithmic chart depicting the value of one Goldmark in Papiermarks during the German hyperinflation in the 1920s

decibel is a unit of measurement associated with logarithmic-scale quantities. It is based on the common logarithm of ratios—10 times the common logarithm of a power ratio or 20 times the common logarithm of a voltage ratio. It is used to quantify the loss of voltage levels in transmitting electrical signals,[66] to describe power levels of sounds in acoustics,[67] and the absorbance of light in the fields of spectrometry and optics. The signal-to-noise ratio describing the amount of unwanted noise in relation to a (meaningful) signal is also measured in decibels.[68] In a similar vein, the peak signal-to-noise ratio is commonly used to assess the quality of sound and image compression methods using the logarithm.[69]

The strength of an earthquake is measured by taking the common logarithm of the energy emitted at the quake. This is used in the moment magnitude scale or the Richter scale. For example, a 5.0 earthquake releases 32 times ($10^{1.5}$) and a 6.0 releases 1000 times (10^3) the energy of a 4.0.[70] Another logarithmic scale is apparent magnitude. It measures the brightness of stars logarithmically.[71] Yet another example is pH in chemistry; pH is the negative of the common logarithm of the activity of hydronium ions (the form hydrogen ions H+

take in water).[72] The activity of hydronium ions in neutral water is 10^{-7} mol·L^{-1}, hence a pH of 7. Vinegar typically has a pH of about 3. The difference of 4 corresponds to a ratio of 10^4 of the activity, that is, vinegar's hydronium ion activity is about 10^{-3} mol·L^{-1}.

Semilog (log-linear) graphs use the logarithmic scale concept for visualization: one axis, typically the vertical one, is scaled logarithmically. For example, the chart at the right compresses the steep increase from 1 million to 1 trillion to the same space (on the vertical axis) as the increase from 1 to 1 million. In such graphs, exponential functions of the form $f(x) = a \cdot b^x$ appear as straight lines with slope equal to the logarithm of b. Log-log graphs scale both axes logarithmically, which causes functions of the form $f(x) = a \cdot x^k$ to be depicted as straight lines with slope equal to the exponent k. This is applied in visualizing and analyzing power laws.[73]

16.7.2   Psychology

Logarithms occur in several laws describing human perception:[74][75] Hick's law proposes a logarithmic relation between the time individuals take to choose an alternative and the number of choices they have.[76] Fitts's law predicts that the time required to rapidly move to a target area is a logarithmic function of the distance to and the size of the target.[77] In psychophysics, the Weber–Fechner law proposes a logarithmic relationship between stimulus and sensation such as the actual vs. the perceived weight of an item a person is carrying.[78] (This "law", however, is less precise than more recent models, such as the Stevens' power law.[79])

Psychological studies found that individuals with little mathematics education tend to estimate quantities logarithmically, that is, they position a number on an unmarked line according to its logarithm, so that 10 is positioned as close to 100 as 100 is to 1000. Increasing education shifts this to a linear estimate (positioning 1000 10x as far away) in some circumstances, while logarithms are used when the numbers to be plotted are difficult to plot linearly.[80][81]

16.7.3   Probability theory and statistics

Logarithms arise in probability theory: the law of large numbers dictates that, for a fair coin, as the number of coin-tosses increases to infinity, the observed proportion of heads approaches one-half. The fluctuations of this proportion about one-half are described by the law of the iterated logarithm.[82]

Logarithms also occur in log-normal distributions. When the logarithm of a random variable has a normal distribution, the variable is said to have a log-normal distribution.[83] Log-normal distributions are encountered in many fields, wherever a variable is formed as the product of many independent positive random variables, for example in the study of turbulence.[84]

Logarithms are used for maximum-likelihood estimation of parametric statistical models. For such a model, the likelihood function depends on at least one parameter that must be estimated. A maximum of the likelihood function occurs at the same parameter-value as a maximum of the logarithm of the likelihood (the "*log likelihood*"), because the logarithm is an increasing function. The log-likelihood is easier to maximize, especially for the multiplied likelihoods for independent random variables.[85]

Benford's law describes the occurrence of digits in many data sets, such as heights of buildings. According to Benford's law, the probability that the first decimal-digit of an item in the data sample is d (from 1 to 9) equals $\log_{10}(d + 1) - \log_{10}(d)$, *regardless* of the unit of measurement.[86] Thus, about 30% of the data can be expected to have 1 as first digit, 18% start with 2, etc. Auditors examine deviations from Benford's law to detect fraudulent accounting.[87]

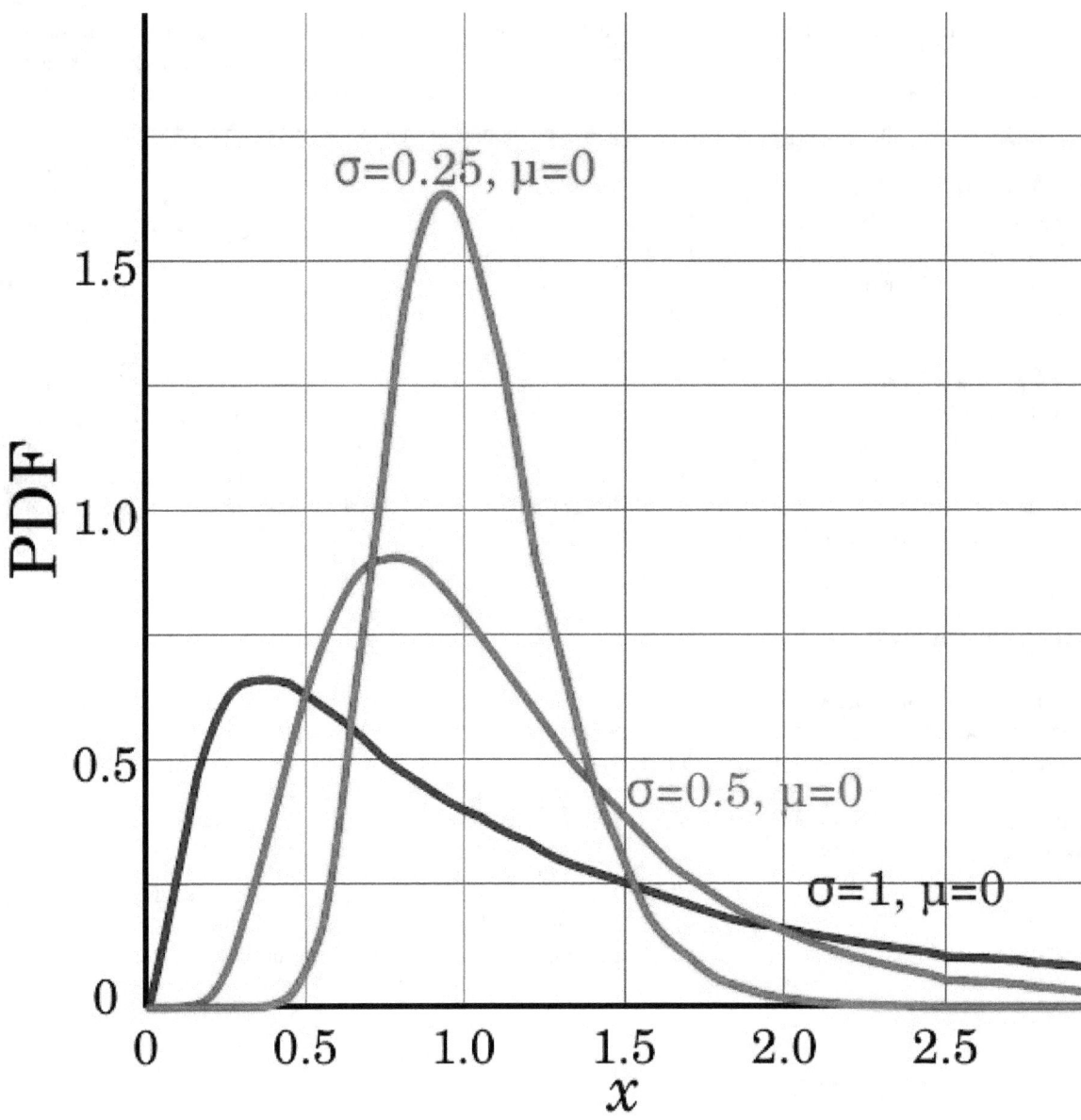

Three probability density functions (PDF) of random variables with log-normal distributions. The location parameter μ, which is zero for all three of the PDFs shown, is the mean of the logarithm of the random variable, not the mean of the variable itself.

16.7.4 Computational complexity

Analysis of algorithms is a branch of computer science that studies the performance of algorithms (computer programs solving a certain problem).[88] Logarithms are valuable for describing algorithms that divide a problem into smaller ones, and join the solutions of the subproblems.[89]

For example, to find a number in a sorted list, the binary search algorithm checks the middle entry and proceeds with the half before or after the middle entry if the number is still not found. This algorithm requires, on average, $\log_2(N)$ comparisons, where N is the list's length.[90] Similarly, the merge sort algorithm sorts an unsorted list by dividing the list into halves and sorting these first before merging the results. Merge sort algorithms typically require a time approximately proportional to $N \cdot \log(N)$.[91] The base of the logarithm is not specified here, because the result only changes by a constant factor when another base is used. A constant factor is usually disregarded in the analysis of algorithms under the standard uniform cost model.[92]

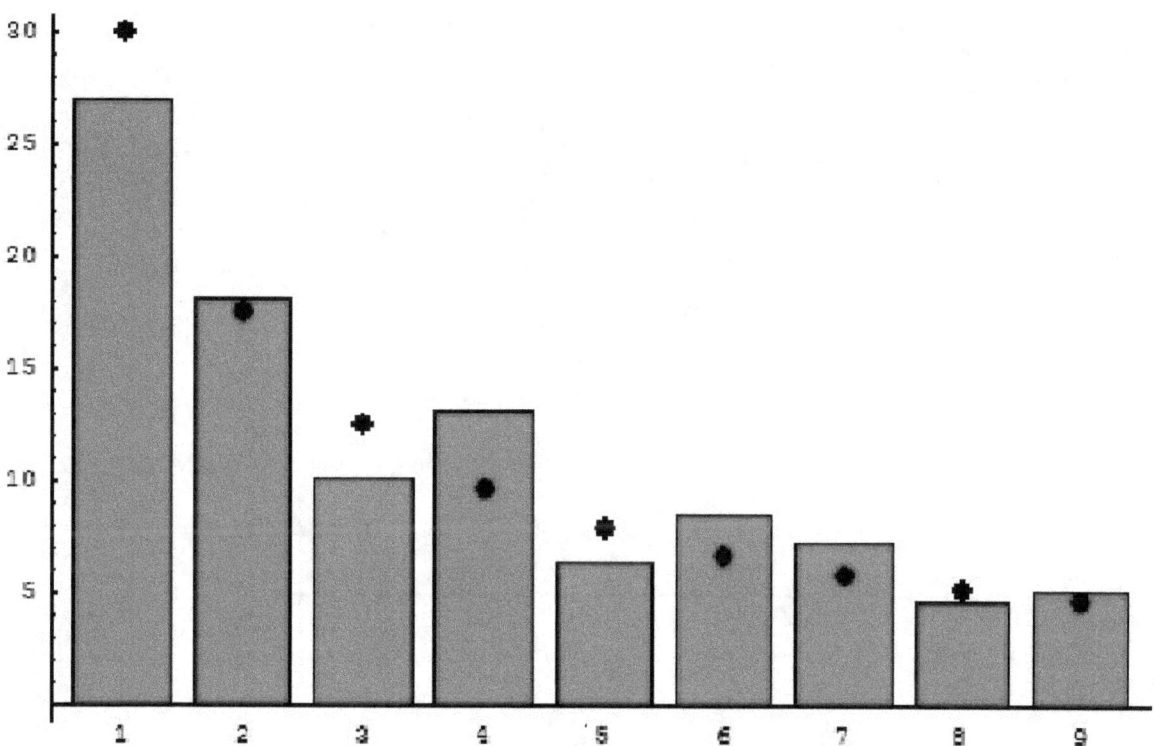

Distribution of first digits (in %, red bars) in the population of the 237 countries of the world. Black dots indicate the distribution predicted by Benford's law.

A function $f(x)$ is said to grow logarithmically if $f(x)$ is (exactly or approximately) proportional to the logarithm of x. (Biological descriptions of organism growth, however, use this term for an exponential function.[93]) For example, any natural number N can be represented in binary form in no more than $\log_2(N) + 1$ bits. In other words, the amount of memory needed to store N grows logarithmically with N.

16.7.5   Entropy and chaos

Entropy is broadly a measure of the disorder of some system. In statistical thermodynamics, the entropy S of some physical system is defined as

$$S = -k \sum_i p_i \ln(p_i).$$

The sum is over all possible states i of the system in question, such as the positions of gas particles in a container. Moreover, pi is the probability that the state i is attained and k is the Boltzmann constant. Similarly, entropy in information theory measures the quantity of information. If a message recipient may expect any one of N possible messages with equal likelihood, then the amount of information conveyed by any one such message is quantified as $\log_2(N)$ bits.[94]

Lyapunov exponents use logarithms to gauge the degree of chaoticity of a dynamical system. For example, for a particle moving on an oval billiard table, even small changes of the initial conditions result in very different paths of the particle. Such systems are chaotic in a deterministic way, because small measurement errors of the initial state predictably lead to largely different final states.[95] At least one Lyapunov exponent of a deterministically chaotic system is positive.

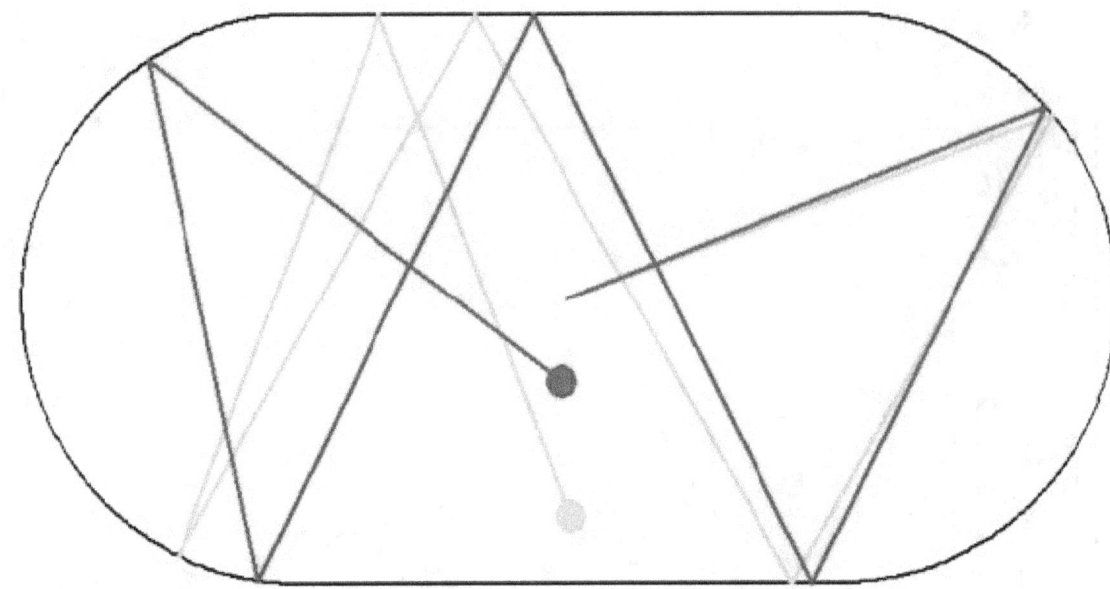

Billiards on an oval billiard table. Two particles, starting at the center with an angle differing by one degree, take paths that diverge chaotically because of reflections at the boundary.

The Sierpinski triangle (at the right) is constructed by repeatedly replacing equilateral triangles by three smaller ones.

16.7.6 Fractals

Logarithms occur in definitions of the dimension of fractals.[96] Fractals are geometric objects that are self-similar: small parts reproduce, at least roughly, the entire global structure. The Sierpinski triangle (pictured) can be covered by three copies of itself, each having sides half the original length. This makes the Hausdorff dimension of this structure $\ln(3)/\ln(2)$ ≈ 1.58. Another logarithm-based notion of dimension is obtained by counting the number of boxes needed to cover the fractal in question.

16.7.7 Music

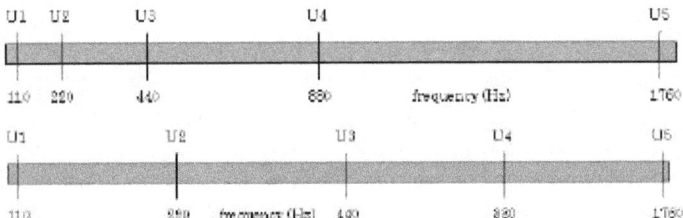

Four different octaves shown on a linear scale, then shown on a logarithmic scale (as the ear hears them).

Logarithms are related to musical tones and intervals. In equal temperament, the frequency ratio depends only on the interval between two tones, not on the specific frequency, or pitch, of the individual tones. For example, the note *A* has a frequency of 440 Hz and *B-flat* has a frequency of 466 Hz. The interval between *A* and *B-flat* is a semitone, as is the one between *B-flat* and *B* (frequency 493 Hz). Accordingly, the frequency ratios agree:

$$\frac{466}{440} \approx \frac{493}{466} \approx 1.059 \approx \sqrt[12]{2}.$$

Therefore, logarithms can be used to describe the intervals: an interval is measured in semitones by taking the base-$2^{1/12}$ logarithm of the frequency ratio, while the base-$2^{1/1200}$ logarithm of the frequency ratio expresses the interval in cents, hundredths of a semitone. The latter is used for finer encoding, as it is needed for non-equal temperaments.[97]

16.7.8   Number theory

Natural logarithms are closely linked to counting prime numbers (2, 3, 5, 7, 11, ...), an important topic in number theory. For any integer x, the quantity of prime numbers less than or equal to x is denoted $\pi(x)$. The prime number theorem asserts that $\pi(x)$ is approximately given by

$$\frac{x}{\ln(x)},$$

in the sense that the ratio of $\pi(x)$ and that fraction approaches 1 when x tends to infinity.[98] As a consequence, the probability that a randomly chosen number between 1 and x is prime is inversely proportional to the number of decimal digits of x. A far better estimate of $\pi(x)$ is given by the offset logarithmic integral function Li(x), defined by

$$\mathrm{Li}(x) = \int_2^x \frac{1}{\ln(t)}\, dt.$$

The Riemann hypothesis, one of the oldest open mathematical conjectures, can be stated in terms of comparing $\pi(x)$ and Li(x).[99] The Erdős–Kac theorem describing the number of distinct prime factors also involves the natural logarithm.

The logarithm of n factorial, $n! = 1 \cdot 2 \cdot \ldots \cdot n$, is given by

$$\ln(n!) = \ln(1) + \ln(2) + \cdots + \ln(n).$$

This can be used to obtain Stirling's formula, an approximation of $n!$ for large n.[100]

16.8   Generalizations

16.8.1   Complex logarithm

Main article: Complex logarithm
The complex numbers a solving the equation

$$e^a = z.$$

are called *complex logarithms*. Here, z is a complex number. A complex number is commonly represented as $z = x + iy$, where x and y are real numbers and i is the imaginary unit. Such a number can be visualized by a point in the complex plane, as shown at the right. The polar form encodes a non-zero complex number z by its absolute value, that is, the distance r to the origin, and an angle between the x axis and the line passing through the origin and z. This angle is called the argument of z. The absolute value r of z is

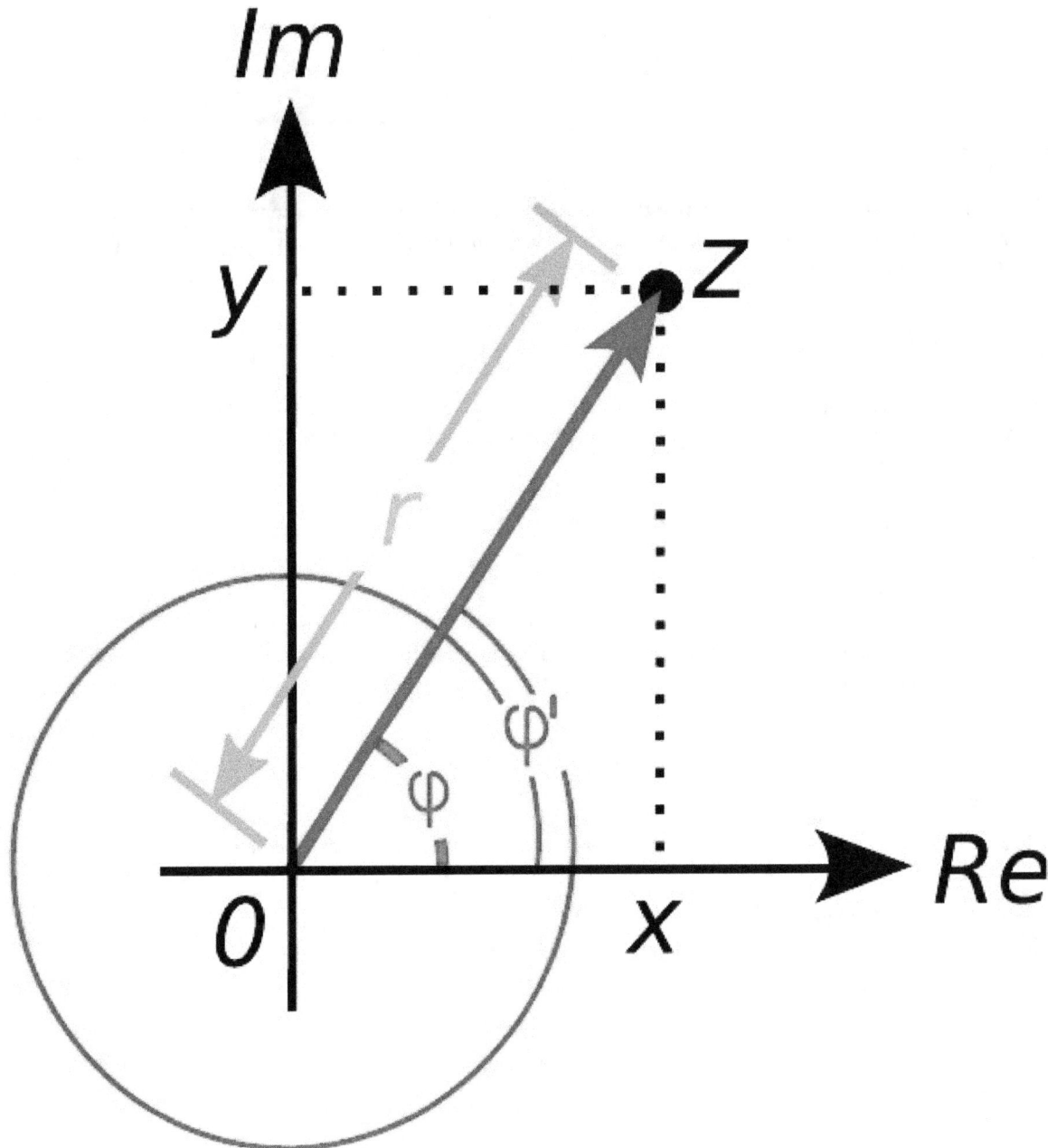

Polar form of z = x + iy. Both φ and φ' are arguments of z.

$$r = \sqrt{x^2 + y^2}.$$

The argument is not uniquely specified by z: both φ and φ' = φ + 2π are arguments of z because adding 2π radians or 360 degrees[nb 6] to φ corresponds to "winding" around the origin counter-clock-wise by a turn. The resulting complex number is again z, as illustrated at the right. However, exactly one argument φ satisfies −π < φ and φ ≤ π. It is called the *principal argument*, denoted Arg(z), with a capital A.[101] (An alternative normalization is 0 ≤ Arg(z) < 2π.[102])

Using trigonometric functions sine and cosine, or the complex exponential, respectively, r and φ are such that the following identities hold:[103]

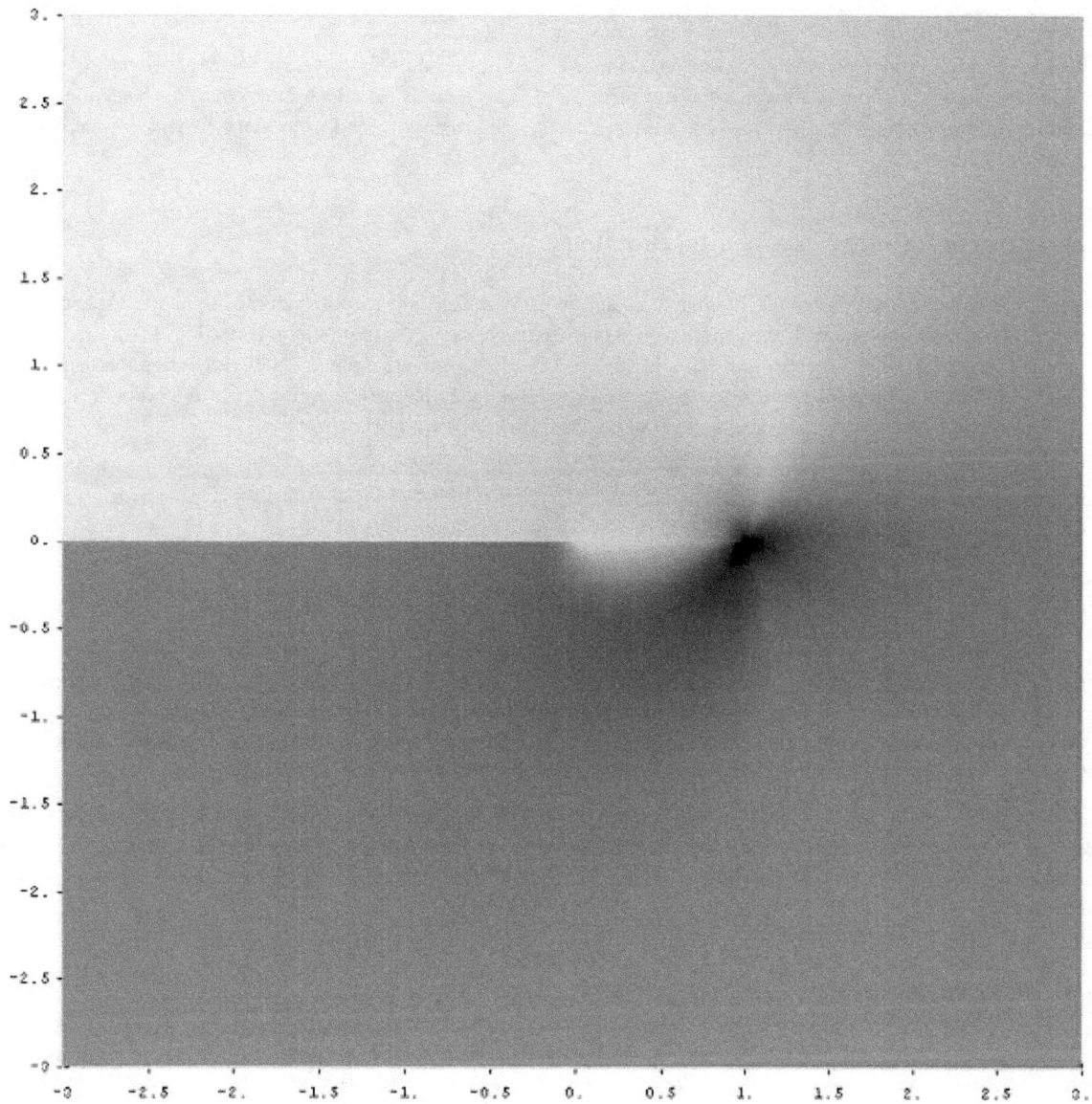

The principal branch of the complex logarithm, Log(z). The black point at z = 1 corresponds to absolute value zero and brighter (more saturated) colors refer to bigger absolute values. The hue of the color encodes the argument of Log(z).

$$z = r(\cos\varphi + i\sin\varphi)$$
$$= re^{i\varphi}.$$

This implies that the a-th power of e equals z, where

$$a = \ln(r) + i(\varphi + 2n\pi).$$

φ is the principal argument $\mathrm{Arg}(z)$ and n is an arbitrary integer. Any such a is called a complex logarithm of z. There are infinitely many of them, in contrast to the uniquely defined real logarithm. If $n = 0$, a is called the *principal value* of the logarithm, denoted $\mathrm{Log}(z)$. The principal argument of any positive real number x is 0; hence $\mathrm{Log}(x)$ is a real number and equals the real (natural) logarithm. However, the above formulas for logarithms of products and powers do *not* generalize to the principal value of the complex logarithm.[104]

The illustration at the right depicts Log(z). The discontinuity, that is, the jump in the hue at the negative part of the x- or real axis, is caused by the jump of the principal argument there. This locus is called a branch cut. This behavior can only be circumvented by dropping the range restriction on φ. Then the argument of z and, consequently, its logarithm become multi-valued functions.

16.8.2   Inverses of other exponential functions

Exponentiation occurs in many areas of mathematics and its inverse function is often referred to as the logarithm. For example, the logarithm of a matrix is the (multi-valued) inverse function of the matrix exponential.[105] Another example is the p-adic logarithm, the inverse function of the p-adic exponential. Both are defined via Taylor series analogous to the real case.[106] In the context of differential geometry, the exponential map maps the tangent space at a point of a manifold to a neighborhood of that point. Its inverse is also called the logarithmic (or log) map.[107]

In the context of finite groups exponentiation is given by repeatedly multiplying one group element b with itself. The discrete logarithm is the integer n solving the equation

$$b^n = x,$$

where x is an element of the group. Carrying out the exponentiation can be done efficiently, but the discrete logarithm is believed to be very hard to calculate in some groups. This asymmetry has important applications in public key cryptography, such as for example in the Diffie–Hellman key exchange, a routine that allows secure exchanges of cryptographic keys over unsecured information channels.[108] Zech's logarithm is related to the discrete logarithm in the multiplicative group of non-zero elements of a finite field.[109]

Further logarithm-like inverse functions include the *double logarithm* ln(ln(x)), the *super- or hyper-4-logarithm* (a slight variation of which is called iterated logarithm in computer science), the Lambert W function, and the logit. They are the inverse functions of the double exponential function, tetration, of $f(w) = we^w$,[110] and of the logistic function, respectively.[111]

16.8.3   Related concepts

From the perspective of group theory, the identity log(cd) = log(c) + log(d) expresses a group isomorphism between positive reals under multiplication and reals under addition. Logarithmic functions are the only continuous isomorphisms between these groups.[112] By means of that isomorphism, the Haar measure (Lebesgue measure) dx on the reals corresponds to the Haar measure dx/x on the positive reals.[113] In complex analysis and algebraic geometry, differential forms of the form df/f are known as forms with logarithmic poles.[114]

The polylogarithm is the function defined by

$$\mathrm{Li}_s(z) = \sum_{k=1}^{\infty} \frac{z^k}{k^s}.$$

It is related to the natural logarithm by $\mathrm{Li}_1(z) = -\ln(1 - z)$. Moreover, Li$s$(1) equals the Riemann zeta function $\zeta(s)$.[115]

16.9   See also

- Cologarithm
- Exponential function

- Decimal exponent (dex)

- Index of logarithm articles

16.10   Notes

[1] The restrictions on *x* and *b* are explained in the section "Analytic properties".

[2] Some mathematicians disapprove of this notation. In his 1985 autobiography, Paul Halmos criticized what he considered the "childish ln notation," which he said no mathematician had ever used.[13] The notation was invented by Irving Stringham, a mathematician.[14][15]

[3] For example C, Java, Haskell, and BASIC.

[4] The same series holds for the principal value of the complex logarithm for complex numbers z satisfying $|z - 1| < 1$.

[5] The same series holds for the principal value of the complex logarithm for complex numbers z with positive real part.

[6] See radian for the conversion between 2π and 360 degrees.

16.11   References

[1] Shirali, Shailesh (2002), *A Primer on Logarithms*, Hyderabad: Universities Press, ISBN 978-81-7371-414-6, esp. section 2

[2] Kate, S.K.; Bhapkar, H.R. (2009), *Basics Of Mathematics*, Pune: Technical Publications, ISBN 978-81-8431-755-8, chapter 1

[3] All statements in this section can be found in Shailesh Shirali 2002, section 4, (Douglas Downing 2003, p. 275), or Kate & Bhapkar 2009, p. 1-1, for example.

[4] Bernstein, Stephen; Bernstein, Ruth (1999), *Schaum's outline of theory and problems of elements of statistics. I, Descriptive statistics and probability*, Schaum's outline series, New York: McGraw-Hill, ISBN 978-0-07-005023-5, p. 21

[5] Downing, Douglas (2003), *Algebra the Easy Way*, Barron's Educational Series, Hauppauge, N.Y.: Barron's, ISBN 978-0-7641-1972-9, chapter 17, p. 275

[6] Wegener, Ingo (2005), *Complexity theory: exploring the limits of efficient algorithms*, Berlin, New York: Springer-Verlag, ISBN 978-3-540-21045-0, p. 20

[7] Van der Lubbe, Jan C. A. (1997), *Information Theory*, Cambridge University Press, p. 3, ISBN 9780521467605.

[8] Allen, Elizabeth; Triantaphillidou, Sophie (2011), *The Manual of Photography*, Taylor & Francis, p. 228, ISBN 9780240520377.

[9] Franz Embacher; Petra Oberhuemer, *Mathematisches Lexikon* (in German), mathe online: für Schule, Fachhochschule, Universität unde Selbststudium, retrieved 2011-03-22

[10] Taylor, B. N. (1995), *Guide for the Use of the International System of Units (SI)*, US Department of Commerce

[11] Gullberg, Jan (1997), *Mathematics: from the birth of numbers.*, New York: W. W. Norton & Co, ISBN 978-0-393-04002-9

[12] See footnote 1 in Perl, Yehoshua; Reingold, Edward M. (December 1977). "Understanding the complexity of interpolation search". *Information Processing Letters* **6** (6): 219–222. doi:10.1016/0020-0190(77)90072-2.

[13] Paul Halmos (1985), *I Want to Be a Mathematician: An Automathography*, Berlin, New York: Springer-Verlag, ISBN 978-0-387-96078-4

[14] Irving Stringham (1893), *Uniplanar algebra: being part I of a propædeutic to the higher mathematical analysis*, The Berkeley Press, p. xiii

[15] Roy S. Freedman (2006), *Introduction to Financial Technology*, Amsterdam: Academic Press, p. 59, ISBN 978-0-12-370478-8

[16] See Theorem 3.29 in Rudin, Walter (1984). *Principles of mathematical analysis* (3rd ed., International student ed. ed.). Auckland: McGraw-Hill International. ISBN 978-0070856134.

[17] McFarland, David (2007), *Quarter Tables Revisited: Earlier Tables, Division of Labor in Table Construction, and Later Implementations in Analog Computers*, p. 1

[18] Robson, Eleanor (2008), *Mathematics in Ancient Iraq: A Social History*, p. 227, ISBN 978-0691091822.

[19] Gupta, R. C. (2000), "History of Mathematics in India", in Hoiberg, Dale; Ramchandani, Indu, *Students' Britannica India: Select essays*, Popular Prakashan, p. 329

[20] Stifelio, Michaele (1544), *Arithmetica Integra*, Nuremberg: Iohan Petreium

[21] Bukhshtab, A.A.; Pechaev, V.I. (2001), "Arithmetic", in Hazewinkel, Michiel, *Encyclopedia of Mathematics*, Springer, ISBN 978-1-55608-010-4

[22] Vivian Shaw Groza and Susanne M. Shelley (1972), *Precalculus mathematics*, New York: Holt, Rinehart and Winston, p. 182, ISBN 978-0-03-077670-0

[23] Napier, John (1614), *Mirifici Logarithmorum Canonis Descriptio* [*The Description of the Wonderful Rule of Logarithms*] (in Latin), Edinburgh, Scotland: Andrew Hart

[24] Hobson, Ernest William (1914), *John Napier and the invention of logarithms, 1614*, Cambridge: The University Press

[25] Bürgi, Jost (1620), *Arithmetische und Geometrische Progress Tabulen ...* [*Arithmetic and Geometric Progression Tables ...*] (in German), Prague, (Czech Republic): University [of Prague] Press
Unfortunately, Bürgi did not include, with his table of powers, instructions for using the table. That was published separately. The contents of that publication were reproduced in: Gieswald, Hermann Robert (1856), *Justus Byrg als Mathematiker, und dessen Einleitung zu seinen Logarithmen* [*Justus Byrg as a mathematician, and an introduction to his logarithms*] (in German), Danzig, Prussia: St. Johannisschule, pp. 26 ff.

[26] Boyer 1991, Chapter 14, section "Jobst Bürgi"

[27] Gladstone-Millar, Lynne (2003), *John Napier: Logarithm John*, National Museums Of Scotland, ISBN 978-1-901663-70-9, p. 44

[28] Napier, Mark (1834), *Memoirs of John Napier of Merchiston*, Edinburgh: William Blackwood, p. 392.

[29] William Harrison De Puy (1893), *The Encyclopædia Britannica: a dictionary of arts, sciences, and general literature ; the R.S. Peale reprint*, **17** (9th ed.), Werner Co., p. 179

[30] Maor, Eli (2009), *e: The Story of a Number*, Princeton University Press, ISBN 978-0-691-14134-3, section 2

[31] In 1647, Gregoire de Saint-Vincent published his book, *Opus geometricum quadraturae circuli et sectionum coni* (Geometric work of squaring the circle and conic sections), vol. 2 (Antwerp, (Belgium): Johannes and Jakob Meursius, 1647). On page 586, Proposition CIX, he proves that if the abscissas of points are in geometric proportion, then the areas between a hyperbola and the abscissas are in arithmetic proportion. This finding allowed Saint-Vincent's former student, Alphonse Antonio de Sarasa, to prove that the area between a hyperbola and the abscissa of a point is proportional to the abscissa's logarithm, thus uniting the algebra of logarithms with the geometry of hyperbolas. See: Alphonse Antonio de Sarasa, *Solutio problematis a R.P. Marino Mersenne Minimo propositi ...* [Solution to a problem proposed by the reverend father Marin Mersenne, member of the Minim order ...]. (Antwerp, (Belgium): Johannes and Jakob Meursius, 1649). Sarasa's critical finding occurs on page 16 (near the bottom of the page), where he states: *"Unde hae superficies supplere possunt locum logarithmorum datorum ... "* (Whence these areas can fill the place of the given logarithms ...). [In other words, the areas are proportional to the logarithms.]
See also: Enrique A. González-Velasco, *Journey through Mathematics: Creative Episodes in Its History* (New York, New York: Springer, 2011), page 118.

[32] Alphonse Antonio de Sarasa, *Solutio problematis a R.P. Marino Mersenne Minimo propositi ...* [Solution to a problem proposed by the reverend father Marin Mersenne, member of the Minim order ...]. (Antwerp, (Belgium): Johannes and Jakob Meursius, 1649).
Sarasa realized that given a hyperbola and a pair of points along the abscissa which were related by a geometric progression, then if the abscissas of the points were multiplied together, the abscissa of their product had an area under the hyperbola which equaled the sum of the points' areas under the hyperbola. That is, the logarithm of an abscissa was proportional to the area, under a hyperbola, corresponding to that abscissa. This finding united the algebra of logarithms with the geometry of hyperbolic curves.

- Sarasa's critical finding occurs on page 16 (near the bottom of the page), where he states: *"Unde hae superficies supplere possant locum logarithmorum datorum ... "* (Whence these areas can fill the place of the given logarithms ...). [In other words, the areas are proportional to the logarithms.]

- See also: Enrique A. González-Velasco, *Journey through Mathematics: Creative Episodes in Its History* (New York, New York: Springer, 2011), pp. 119-120.

[33] J. J. O'Connor; E. F. Robertson (September 2001), *The number e*, The MacTutor History of Mathematics archive, retrieved 2009-02-02

[34] Cajori, Florian (1991), *A History of Mathematics* (5th ed.), Providence, RI: AMS Bookstore, ISBN 978-0-8218-2102-2, p. 152

[35] Maor 2009, sections 1, 13

[36] Eves, Howard Whitley (1992), *An introduction to the history of mathematics*, The Saunders series (6th ed.), Philadelphia: Saunders, ISBN 978-0-03-029558-4, section 9-3

[37] Boyer, Carl B. (1991), *A History of Mathematics*, New York: John Wiley & Sons, ISBN 978-0-471-54397-8, p. 484, 489

[38] Bryant, Walter W., *A History of Astronomy*, London: Methuen & Co, p. 44

[39] Campbell-Kelly, Martin (2003), *The history of mathematical tables: from Sumer to spreadsheets*, Oxford scholarship online, Oxford University Press, ISBN 978-0-19-850841-0, section 2

[40] Abramowitz, Milton; Stegun, Irene A., eds. (1972), *Handbook of Mathematical Functions with Formulas, Graphs, and Mathematical Tables* (10th ed.), New York: Dover Publications, ISBN 978-0-486-61272-0, section 4.7., p. 89

[41] Spiegel, Murray R.; Moyer, R.E. (2006), *Schaum's outline of college algebra*, Schaum's outline series, New York: McGraw-Hill, ISBN 978-0-07-145227-4, p. 264

[42] Devlin, Keith (2004). *Sets, functions, and logic: an introduction to abstract mathematics*. Chapman & Hall/CRC mathematics (3rd ed.). Boca Raton, Fla: Chapman & Hall/CRC. ISBN 1-58488-449-5., or see the references in function

[43] Lang, Serge (1997), *Undergraduate analysis*, Undergraduate Texts in Mathematics (2nd ed.), Berlin, New York: Springer-Verlag, ISBN 978-0-387-94841-6, MR 1476913, section III.3

[44] Lang 1997, section IV.2

[45] Dieudonné, Jean (1969). *Foundations of Modern Analysis* 1. Academic Press. p. 84. item (4.3.1)

[46] Stewart, James (2007), *Single Variable Calculus: Early Transcendentals*, Belmont: Thomson Brooks/Cole, ISBN 978-0-495-01169-9, section 1.6

[47] "Calculation of $d/dx(Log(b,x))$". Wolfram Alpha. Wolfram Research. Retrieved 15 March 2011.

[48] Kline, Morris (1998), *Calculus: an intuitive and physical approach*, Dover books on mathematics, New York: Dover Publications, ISBN 978-0-486-40453-0, p. 386

[49] "Calculation of $Integrate(ln(x))$". Wolfram Alpha. Wolfram Research. Retrieved 15 March 2011.

[50] Abramowitz & Stegun, eds. 1972, p. 69

[51] Courant, Richard (1988). *Differential and integral calculus. Vol. 1*, Wiley Classics Library, New York: John Wiley & Sons, ISBN 978-0-471-60842-4, MR 1009558, section III.6

[52] Havil, Julian (2003), *Gamma: Exploring Euler's Constant*, Princeton University Press, ISBN 978-0-691-09983-5, sections 11.5 and 13.8

[53] Nomizu, Katsumi (1996), *Selected papers on number theory and algebraic geometry* 172, Providence, RI: AMS Bookstore, p. 21, ISBN 978-0-8218-0445-2

[54] Baker, Alan (1975), *Transcendental number theory*, Cambridge University Press, ISBN 978-0-521-20461-3, p. 10

[55] Muller, Jean-Michel (2006), *Elementary functions* (2nd ed.), Boston, MA: Birkhäuser Boston, ISBN 978-0-8176-4372-0, sections 4.2.2 (p. 72) and 5.5.2 (p. 95)

[56] Hart, Cheney, Lawson; et al. (1968), *Computer Approximations*, SIAM Series in Applied Mathematics, New York: John Wiley, section 6.3, p. 105–111

[57] Zhang, M.; Delgado-Frias, J.G.; Vassiliadis, S. (1994), "Table driven Newton scheme for high precision logarithm generation", *IEE Proceedings Computers & Digital Techniques* **141** (5): 281–292, doi:10.1049/ip-cdt:19941268, ISSN 1350-2387, section 1 for an overview

[58] Meggitt, J. E. (April 1962), "Pseudo Division and Pseudo Multiplication Processes", *IBM Journal*, doi:10.1147/rd.62.0210

[59] Kahan, W. (May 20, 2001), *Pseudo-Division Algorithms for Floating-Point Logarithms and Exponentials*

[60] Abramowitz & Stegun, eds. 1972, p. 68

[61] Sasaki, T.; Kanada, Y. (1982), "Practically fast multiple-precision evaluation of log(x)", *Journal of Information Processing* **5** (4): 247–250, retrieved 30 March 2011

[62] Ahrendt, Timm (1999), *Fast computations of the exponential function*, Lecture notes in computer science **1564**, Berlin, New York: Springer, pp. 302–312, doi:10.1007/3-540-49116-3_28

[63] Maor 2009, p. 135

[64] Frey, Bruce (2006), *Statistics hacks*, Hacks Series, Sebastopol, CA: O'Reilly, ISBN 978-0-596-10164-0, chapter 6, section 64

[65] Ricciardi, Luigi M. (1990), *Lectures in applied mathematics and informatics*, Manchester: Manchester University Press, ISBN 978-0-7190-2671-3, p. 21, section 1.3.2

[66] Bakshi, U. A. (2009), *Telecommunication Engineering*, Pune: Technical Publications, ISBN 978-81-8431-725-1, section 5.2

[67] Maling, George C. (2007), "Noise", in Rossing, Thomas D., *Springer handbook of acoustics*, Berlin, New York: Springer-Verlag, ISBN 978-0-387-30446-5, section 23.0.2

[68] Tashev, Ivan Jelev (2009), *Sound Capture and Processing: Practical Approaches*, New York: John Wiley & Sons, ISBN 978-0-470-31983-3, p. 48

[69] Chui, C.K. (1997), *Wavelets: a mathematical tool for signal processing*, SIAM monographs on mathematical modeling and computation, Philadelphia: Society for Industrial and Applied Mathematics, ISBN 978-0-89871-384-8, p. 180

[70] Crauder, Bruce; Evans, Benny; Noell, Alan (2008), *Functions and Change: A Modeling Approach to College Algebra* (4th ed.), Boston: Cengage Learning, ISBN 978-0-547-15669-9, section 4.4.

[71] Bradt, Hale (2004), *Astronomy methods: a physical approach to astronomical observations*, Cambridge Planetary Science, Cambridge University Press, ISBN 978-0-521-53551-9, section 8.3, p. 231

[72] IUPAC (1997), A. D. McNaught, A. Wilkinson, ed., *Compendium of Chemical Terminology ("Gold Book")* (2nd ed.), Oxford: Blackwell Scientific Publications, doi:10.1351/goldbook, ISBN 978-0-9678550-9-7

[73] Bird, J. O. (2001), *Newnes engineering mathematics pocket book* (3rd ed.), Oxford: Newnes, ISBN 978-0-7506-4992-6, section 34

[74] Goldstein, E. Bruce (2009), *Encyclopedia of Perception*, Encyclopedia of Perception, Thousand Oaks, CA: Sage, ISBN 978-1-4129-4081-8, p. 355–356

[75] Matthews, Gerald (2000), *Human performance: cognition, stress, and individual differences*, Human Performance: Cognition, Stress, and Individual Differences, Hove: Psychology Press, ISBN 978-0-415-04406-6, p. 48

[76] Welford, A. T. (1968), *Fundamentals of skill*, London: Methuen, ISBN 978-0-416-03000-6, OCLC 219156, p. 61

[77] Paul M. Fitts (June 1954), "The information capacity of the human motor system in controlling the amplitude of movement", *Journal of Experimental Psychology* **47** (6): 381–391, doi:10.1037/h0055392, PMID 13174710, reprinted in Paul M. Fitts (1992), "The information capacity of the human motor system in controlling the amplitude of movement" (PDF), *Journal of Experimental Psychology: General* **121** (3): 262–269, doi:10.1037/0096-3445.121.3.262, PMID 1402698, retrieved 30 March 2011

[78] Banerjee, J. C. (1994), *Encyclopaedic dictionary of psychological terms*, New Delhi: M.D. Publications, ISBN 978-81-85880-28-0, OCLC 33860167, p. 304

[79] Nadel, Lynn (2005), *Encyclopedia of cognitive science*, New York: John Wiley & Sons, ISBN 978-0-470-01619-0, lemmas *Psychophysics* and *Perception: Overview*

[80] Siegler, Robert S.; Opfer, John E. (2003), "The Development of Numerical Estimation. Evidence for Multiple Representations of Numerical Quantity" (PDF), *Psychological Science* **14** (3): 237–43, doi:10.1111/1467-9280.02438, PMID 12741747

[81] Dehaene, Stanislas; Izard, Véronique; Spelke, Elizabeth; Pica, Pierre (2008), "Log or Linear? Distinct Intuitions of the Number Scale in Western and Amazonian Indigene Cultures", *Science* **320** (5880): 1217–1220, doi:10.1126/science.1156540, PMC 2610411, PMID 18511690

[82] Breiman, Leo (1992), *Probability*, Classics in applied mathematics, Philadelphia: Society for Industrial and Applied Mathematics, ISBN 978-0-89871-296-4, section 12.9

[83] Aitchison, J.; Brown, J. A. C. (1969), *The lognormal distribution*, Cambridge University Press, ISBN 978-0-521-04011-2, OCLC 301100935

[84] Jean Mathieu and Julian Scott (2000), *An introduction to turbulent flow*, Cambridge University Press, p. 50, ISBN 978-0-521-77538-0

[85] Rose, Colin; Smith, Murray D. (2002), *Mathematical statistics with Mathematica*, Springer texts in statistics, Berlin, New York: Springer-Verlag, ISBN 978-0-387-95234-5, section 11.3

[86] Tabachnikov, Serge (2005), *Geometry and Billiards*, Providence, R.I.: American Mathematical Society, pp. 36–40, ISBN 978-0-8218-3919-5, section 2.1

[87] Durtschi, Cindy; Hillison, William; Pacini, Carl (2004), "The Effective Use of Benford's Law in Detecting Fraud in Accounting Data" (PDF), *Journal of Forensic Accounting* **V**: 17–34

[88] Wegener, Ingo (2005), *Complexity theory: exploring the limits of efficient algorithms*, Berlin, New York: Springer-Verlag, ISBN 978-3-540-21045-0, pages 1-2

[89] Harel, David; Feldman, Yishai A. (2004), *Algorithmics: the spirit of computing*, New York: Addison-Wesley, ISBN 978-0-321-11784-7, p. 143

[90] Knuth, Donald (1998), *The Art of Computer Programming*, Reading, Mass.: Addison-Wesley, ISBN 978-0-201-89685-5, section 6.2.1, pp. 409–426

[91] Donald Knuth 1998, section 5.2.4, pp. 158–168

[92] Wegener, Ingo (2005), *Complexity theory: exploring the limits of efficient algorithms*, Berlin, New York: Springer-Verlag, p. 20, ISBN 978-3-540-21045-0

[93] Mohr, Hans; Schopfer, Peter (1995), *Plant physiology*, Berlin, New York: Springer-Verlag, ISBN 978-3-540-58016-4, chapter 19, p. 298

[94] Eco, Umberto (1989), *The open work*, Harvard University Press, ISBN 978-0-674-63976-8, section III.1

[95] Sprott, Julien Clinton (2010), *Elegant Chaos: Algebraically Simple Chaotic Flows*, New Jersey: World Scientific, ISBN 978-981-283-881-0, section 1.9

[96] Helmberg, Gilbert (2007), *Getting acquainted with fractals*, De Gruyter Textbook, Berlin, New York: Walter de Gruyter, ISBN 978-3-11-019092-2

[97] Wright, David (2009), *Mathematics and music*, Providence, RI: AMS Bookstore, ISBN 978-0-8218-4873-9, chapter 5

[98] Bateman, P. T.; Diamond, Harold G. (2004), *Analytic number theory: an introductory course*, New Jersey: World Scientific, ISBN 978-981-256-080-3, OCLC 492669517, theorem 4.1

[99] P. T. Bateman & Diamond 2004, Theorem 8.15

[100] Slomson, Alan B. (1991), *An introduction to combinatorics*, London: CRC Press, ISBN 978-0-412-35370-3, chapter 4

[101] Ganguly, S. (2005), *Elements of Complex Analysis*, Kolkata: Academic Publishers, ISBN 978-81-87504-86-3, Definition 1.6.3

[102] Nevanlinna, Rolf Herman; Paatero, Veikko (2007), "Introduction to complex analysis", *London: Hilger* (Providence, RI: AMS Bookstore), Bibcode:1974aitc.book.....W, ISBN 978-0-8218-4399-4, section 5.9

[103] Moore, Theral Orvis; Hadlock, Edwin H. (1991), *Complex analysis*, Singapore: World Scientific, ISBN 978-981-02-0246-0, section 1.2

[104] Wilde, Ivan Francis (2006), *Lecture notes on complex analysis*, London: Imperial College Press, ISBN 978-1-86094-642-4, theorem 6.1.

[105] Higham, Nicholas (2008), *Functions of Matrices. Theory and Computation*, Philadelphia, PA: SIAM, ISBN 978-0-89871-646-7, chapter 11.

[106] Neukirch, Jürgen (1999), *Algebraic Number Theory*, Grundlehren der mathematischen Wissenschaften **322**, Berlin: Springer-Verlag, ISBN 978-3-540-65399-8, Zbl 0956.11021, MR 1697859, section II.5.

[107] Hancock, Edwin R.; Martin, Ralph R.; Sabin, Malcolm A. (2009), *Mathematics of Surfaces XIII: 13th IMA International Conference York, UK, September 7–9, 2009 Proceedings*, Springer, p. 379, ISBN 978-3-642-03595-1

[108] Stinson, Douglas Robert (2006), *Cryptography: Theory and Practice* (3rd ed.), London: CRC Press, ISBN 978-1-58488-508-5

[109] Lidl, Rudolf; Niederreiter, Harald (1997), *Finite fields*, Cambridge University Press, ISBN 978-0-521-39231-0

[110] Corless, R.; Gonnet, G.; Hare, D.; Jeffrey, D.; Knuth, Donald (1996), "On the Lambert W function" (PDF), *Advances in Computational Mathematics* (Berlin, New York: Springer-Verlag) **5**: 329–359, doi:10.1007/BF02124750, ISSN 1019-7168

[111] Cherkassky, Vladimir; Cherkassky, Vladimir S.; Mulier, Filip (2007), *Learning from data: concepts, theory, and methods*, Wiley series on adaptive and learning systems for signal processing, communications, and control, New York: John Wiley & Sons, ISBN 978-0-471-68182-3, p. 357

[112] Bourbaki, Nicolas (1998), *General topology. Chapters 5—10*, Elements of Mathematics, Berlin, New York: Springer-Verlag, ISBN 978-3-540-64563-4, MR 1726872, section V.4.1

[113] Ambartzumian, R. V. (1990), *Factorization calculus and geometric probability*, Cambridge University Press, ISBN 978-0-521-34535-4, section 1.4

[114] Esnault, Hélène; Viehweg, Eckart (1992), *Lectures on vanishing theorems*, DMV Seminar **20**, Basel, Boston: Birkhäuser Verlag, ISBN 978-3-7643-2822-1, MR 1193913, section 2

[115] Apostol, T.M. (2010), "Logarithm", in Olver, Frank W. J.; Lozier, Daniel M.; Boisvert, Ronald F.; Clark, Charles W., *NIST Handbook of Mathematical Functions*, Cambridge University Press, ISBN 978-0521192255, MR 2723248

16.12   External links

- Media related to Logarithm at Wikimedia Commons

- The dictionary definition of logarithm at Wiktionary

- Khan Academy: Logarithms, free online micro lectures

- Hazewinkel, Michiel, ed. (2001), "Logarithmic function", *Encyclopedia of Mathematics*, Springer, ISBN 978-1-55608-010-4

- Colin Byfleet, *Educational video on logarithms*, retrieved 2010-10-12

- Edward Wright, *Translation of Napier's work on logarithms*, retrieved 2010-10-12

Chapter 17

Ceteris paribus

Ceteris paribus or *caeteris paribus* is a Latin phrase meaning "with other things the same" or "all or **other things being equal** or held constant" or "**all other things being equal**" or "**all else being equal**". A prediction or a statement about a causal, empirical, or logical relation between two states of affairs is *ceteris paribus* if it is acknowledged that the prediction, although usually accurate in expected conditions, can fail or the relation can be abolished by intervening factors.[1]

A *ceteris paribus* **assumption** is often key to scientific inquiry, as scientists seek to screen out factors that perturb a relation of interest. Thus, epidemiologists seek to control independent variables as factors that may influence dependent variables—the outcomes or effects of interest. Likewise, in scientific modeling, simplifying assumptions permit illustration or elucidation of concepts thought relevant within the sphere of inquiry.

Whereas fundamental physics tends to state universal laws, other sciences, such as biology, psychology, and economics, tend to state laws that hold true in "normal conditions" but have exceptions, *ceteris paribus* **laws** (cp laws).[2] The focus on universal laws is a criterion distinguishing fundamental physics as fundamental science, whereas *ceteris paribus* laws are predominant in most other sciences as special sciences, whose laws hold in special cases.[2] This distinction assumes a logical empiricist view of science. It does not readily apply in a mechanistic understanding of scientific discovery. There is reasonable disagreement as to whether mechanisms or laws are the appropriate model, though mechanisms are the favored method. [3]

17.1 Economics

17.1.1 Clause

One of the disciplines in which *ceteris paribus* clauses are most widely used is economics, in which they are employed to simplify the formulation and description of economic outcomes. When using *ceteris paribus* in economics, one assumes that all other variables except those under immediate consideration are held constant. For example, it can be predicted that if the price of beef *increases*—*ceteris paribus*—the quantity of beef demanded by buyers will *decrease*. In this example, the clause is used to operationally describe everything surrounding the relationship between both the *price* and the *quantity demanded* of an ordinary good.

This operational description intentionally ignores both known and unknown factors that may also influence the relationship between price and quantity demanded, and thus to assume *ceteris paribus* is to assume away any interference with the given example. Such factors that would be intentionally ignored include: a change in the price of substitute goods, (e.g., the price of pork or lamb); a change in the level of risk aversion among buyers (e.g., due to an increase in the fear of mad cow disease); and a change in the level of overall demand for a good regardless of its current price (e.g., a societal shift toward vegetarianism).

The clause is often loosely translated as "holding all else constant." It does not imply that no other things will in fact change; rather, it isolates the effect of one particular change. Holding all other things constant is directly analogous to using a partial derivative in calculus rather than a total derivative, and to running a regression containing multiple variables rather than just one in order to isolate the individual effect of one of the variables.

17.1.2 Characterization given by Alfred Marshall

The clause is used to consider the effect of some causes in isolation, by assuming that other influences are absent. Alfred Marshall expressed the use of the clause as follows:

> The element of time is a chief cause of those difficulties in economic investigations which make it necessary for man with his limited powers to go step by step; breaking up a complex question, studying one bit at a time, and at last combining his partial solutions into a more or less complete solution of the whole riddle. In breaking it up, he segregates those disturbing causes, whose wanderings happen to be inconvenient, for the time in a pound called Ceteris Paribus. The study of some group of tendencies is isolated by the assumption other things being equal: the existence of other tendencies is not denied, but their disturbing effect is neglected for a time. The more the issue is thus narrowed, the more exactly can it be handled: but also the less closely does it correspond to real life. Each exact and firm handling of a narrow issue, however, helps towards treating broader issues, in which that narrow issue is contained, more exactly than would otherwise have been possible. With each step more things can be let out of the pound; exact discussions can be made less abstract, realistic discussions can be made less inexact than was possible at an earlier stage. (Principles of Economics, Bk.V,Ch.V in paragraph V.V.10).

17.1.3 Two uses

The above passage by Marshall highlights two ways in which the *ceteris paribus* clause may be used: The one is *hypothetical*, in the sense that some factor is assumed fixed in order to analyse the influence of another factor in isolation. This would be **hypothetical isolation**. An example would be the hypothetical separation of the income effect and the substitution effect of a price change, which actually go together. The other use of the *ceteris paribus* clause is to see it as a means for obtaining an approximate solution. Here it would yield a **substantive isolation**.

Substantive isolation has two aspects: temporal and causal. **Temporal isolation** requires the factors fixed under the *ceteris paribus* clause to actually move so slowly relative to the other influence that they can be taken as practically constant at any point in time. So, if vegetarianism spreads very slowly, inducing a slow decline in the demand for beef, and the market for beef clears comparatively quickly, we can determine the price of beef at any instant by the intersection of supply and demand, and the changing demand for beef will account for the price changes over time (Temporary Equilibrium Method).

The other aspect of substantive isolation is **causal isolation**: those factors frozen under a *ceteris paribus* clause should not significantly be affected by the processes under study. If a change in government policies induces changes in consumers' behaviour on the same time scale, the assumption that consumer behaviour remains unchanged while policy changes is inadmissible as a substantive isolation (Lucas critique).

17.2 See also

- *Mutatis mutandis*

- List of Latin phrases

- Confounding

- Occam's razor

17.3   References

[1] Schlicht, E. (1985). *Isolation and Aggregation in Economics*. Springer Verlag. ISBN 0-387-15254-7. chapter 2

[2] Alexander Reutlinger, Gerhard Schurz & Andreas Hüttemann, "Ceteris paribus laws", in Edward N Zalta, ed, *The Stanford Encyclopedia of Philosophy*, Spring 2014 edn.

[3] Glennan, S. (2014). Mechanisms. In M. Curd & S. Psillos (Eds.), The Routledge companion to philosophy of science (2nd ed., pp. 420–428). New York: Routledge.

- Persky, Joseph (1990). "Retroperspectives: Ceteris Paribus". *The Journal of Economic Perspectives* 4 (2): 187–193. doi:10.1257/jep.4.2.187. JSTOR 1942898.

17.4   External links

- Ceteris paribus laws at PhilPapers

- Ceteris paribus laws at the Indiana Philosophy Ontology Project

- Ceteris paribus laws entry in the *Stanford Encyclopedia of Philosophy*

- Listen to Ceteris Paribus

Chapter 18

Scientific modelling

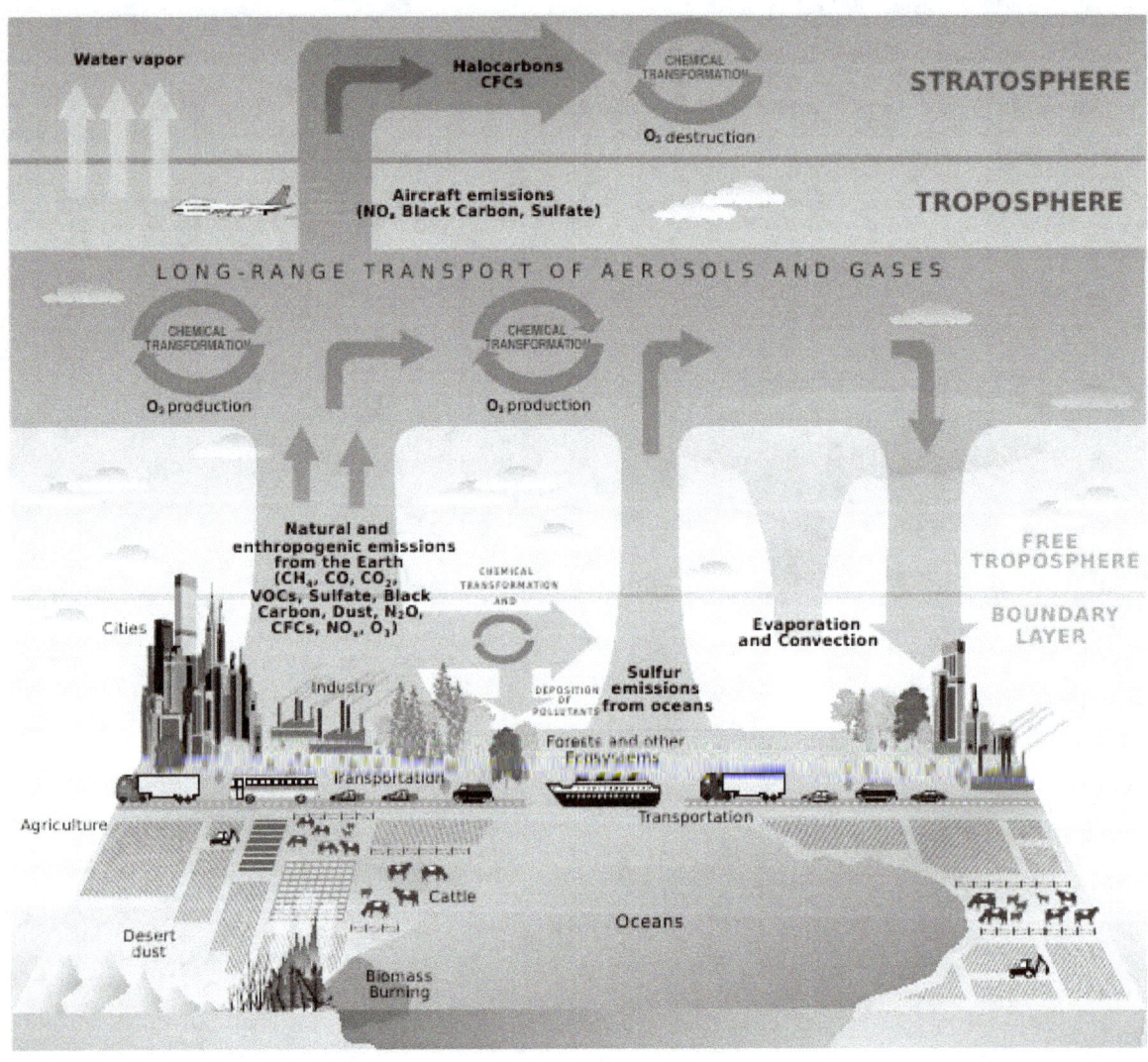

Example of scientific modelling. A schematic of chemical and transport processes related to atmospheric composition.

Scientific modelling is a scientific activity, the aim of which is to make a particular part or feature of the world easier to understand, define, quantify, visualize, or simulate by referencing it to existing and usually commonly accepted knowledge.

It requires selecting and identifying relevant aspects of a situation in the real world and then using different types of models for different aims, such as conceptual models to better understand, operational models to operationalize, mathematical models to quantify, and graphical models to visualize the subject. Modelling is an essential and inseparable part of scientific activity, and many scientific disciplines have their own ideas about specific types of modelling.[1][2]

There is also an increasing attention to scientific modelling[3] in fields such as science education, philosophy of science, systems theory, and knowledge visualization. There is growing collection of methods, techniques and meta-theory about all kinds of specialized scientific modelling.

18.1 Overview

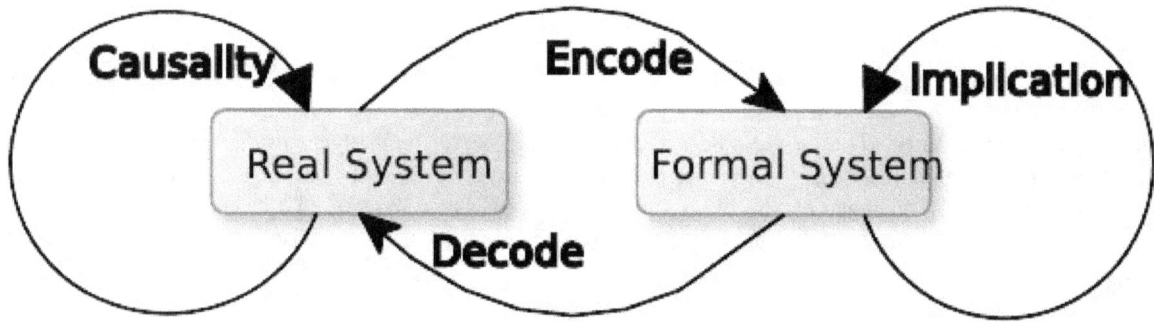

A scientific model seeks to represent empirical objects, phenomena, and physical processes in a logical and objective way. All models are *in simulacra*, that is, simplified reflections of reality that, despite being approximations, can be extremely useful.[4] Building and disputing models is fundamental to the scientific enterprise. Complete and true representation may be impossible, but scientific debate often concerns which is the better model for a given task, e.g., which is the more accurate climate model for seasonal forecasting.[5]

Attempts to formalize the principles of the empirical sciences use an interpretation to model reality, in the same way logicians axiomatize the principles of logic. The aim of these attempts is to construct a formal system that will not produce theoretical consequences that are contrary to what is found in reality. Predictions or other statements drawn from such a formal system mirror or map the real world only insofar as these scientific models are true.[6] [7]

For the scientist, a model is also a way in which the human thought processes can be amplified.[8] For instance, models that are rendered in software allow scientists to leverage computational power to simulate, visualize, manipulate and gain intuition about the entity, phenomenon, or process being represented. Such computer models are *in silico*. Other types of scientific model are *in vivo* (living models, such as laboratory rats) and *in vitro* (in glassware, such as tissue culture).[9]

18.2 Basics of scientific modelling

18.2.1   Modelling as a substitute for direct measurement and experimentation

Models are typically used when it is either impossible or impractical to create experimental conditions in which scientists can directly measure outcomes. Direct measurement of outcomes under controlled conditions (see Scientific method) will always be more reliable than modelled estimates of outcomes.

18.2.2   Simulation

A simulation is the implementation of a model. A steady state simulation provides information about the system at a specific instant in time (usually at equilibrium, if such a state exists). A dynamic simulation provides information over time. A simulation brings a model to life and shows how a particular object or phenomenon will behave. Such a simulation can be useful for testing, analysis, or training in those cases where real-world systems or concepts can be represented by models.[10]

18.2.3 Structure

Structure is a fundamental and sometimes intangible notion covering the recognition, observation, nature, and stability of patterns and relationships of entities. From a child's verbal description of a snowflake, to the detailed scientific analysis of the properties of magnetic fields, the concept of structure is an essential foundation of nearly every mode of inquiry and discovery in science, philosophy, and art.[11]

18.2.4 Systems

A system is a set of interacting or interdependent entities, real or abstract, forming an integrated whole. In general, a system is a construct or collection of different elements that together can produce results not obtainable by the elements alone.[12] The concept of an 'integrated whole' can also be stated in terms of a system embodying a set of relationships which are differentiated from relationships of the set to other elements, and from relationships between an element of the set and elements not a part of the relational regime. There are two types of system models: 1) discrete in which the variables change instantaneously at separate points in time and, 2) continuous where the state variables change continuously with respect to time.[13]

18.2.5 Generating a model

Modelling refers to the process of generating a model as a conceptual representation of some phenomenon. Typically a model will refer only to some aspects of the phenomenon in question, and two models of the same phenomenon may be essentially different, that is to say that the differences between them comprise more than just a simple renaming of components.

Such differences may be due to differing requirements of the model's end users, or to conceptual or aesthetic differences among the modellers and to contingent decisions made during the modelling process. Considerations that may influence the structure of a model might be the modeller's preference for a reduced ontology, preferences regarding statistical models versus deterministic models, discrete versus continuous time, etc. In any case, users of a model need to understand the assumptions made that are pertinent to its validity for a given use.

Building a model requires abstraction. Assumptions are used in modelling in order to specify the domain of application of the model. For example, the special theory of relativity assumes an inertial frame of reference. This assumption was contextualized and further explained by the general theory of relativity. A model makes accurate predictions when its assumptions are valid, and might well not make accurate predictions when its assumptions do not hold. Such assumptions are often the point with which older theories are succeeded by new ones (the general theory of relativity works in non-inertial reference frames as well).

The term "assumption" is actually broader than its standard use, etymologically speaking. The Oxford English Dictionary (OED) and online Wiktionary indicate its Latin source as *assumere* ("accept, to take to oneself, adopt, usurp"), which is a conjunction of *ad-* ("to, towards, at") and *sumere* (to take). The root survives, with shifted meanings, in the Italian *sumere* and Spanish *sumir*. In the OED, "assume" has the senses of (i) "investing oneself with (an attribute), " (ii) "to undertake" (especially in Law), (iii) "to take to oneself in appearance only, to pretend to possess," and (iv) "to suppose a thing to be." Thus, "assumption" connotes other associations than the contemporary standard sense of "that which is assumed or taken for granted: a supposition, postulate," and deserves a broader analysis in the philosophy of science.

18.2.6 Evaluating a model

See also: Models of scientific inquiry § Choice of a theory

A model is evaluated first and foremost by its consistency to empirical data; any model inconsistent with reproducible observations must be modified or rejected. One way to modify the model is by restricting the domain over which it is credited with having high validity. A case in point is Newtonian physics, which is highly useful except for the very small, the very fast, and the very massive phenomena of the universe. However, a fit to empirical data alone is not sufficient for a model to be accepted as valid. Other factors important in evaluating a model include:

- Ability to explain past observations

- Ability to predict future observations

- Cost of use, especially in combination with other models

- Refutability, enabling estimation of the degree of confidence in the model

- Simplicity, or even aesthetic appeal

People may attempt to quantify the evaluation of a model using a utility function.

18.2.7 Visualization

Visualization is any technique for creating images, diagrams, or animations to communicate a message. Visualization through visual imagery has been an effective way to communicate both abstract and concrete ideas since the dawn of man. Examples from history include cave paintings, Egyptian hieroglyphs, Greek geometry, and Leonardo da Vinci's revolutionary methods of technical drawing for engineering and scientific purposes.

18.3   Types of scientific modelling

18.4   Applications

18.4.1   Modelling and Simulation

One application of scientific modelling is the field of "Modelling and Simulation", generally referred to as "M&S". M&S has a spectrum of applications which range from concept development and analysis, through experimentation, measurement and verification, to disposal analysis. Projects and programs may use hundreds of different simulations, simulators and model analysis tools.

The figure shows how Modelling and Simulation is used as a central part of an integrated program in a Defence capability development process.[10]

18.4.2   Model Based Learning in Education

Model Based learning in education, particularly in relation to learning science involves students creating models for scientific concepts in order to [14] :

- Gain insight of the scientific idea(s)

- Acquire deeper understanding of the subject through visualization of the model

- Improve student engagement in the course

Different types of model based learning techniques include [14] :

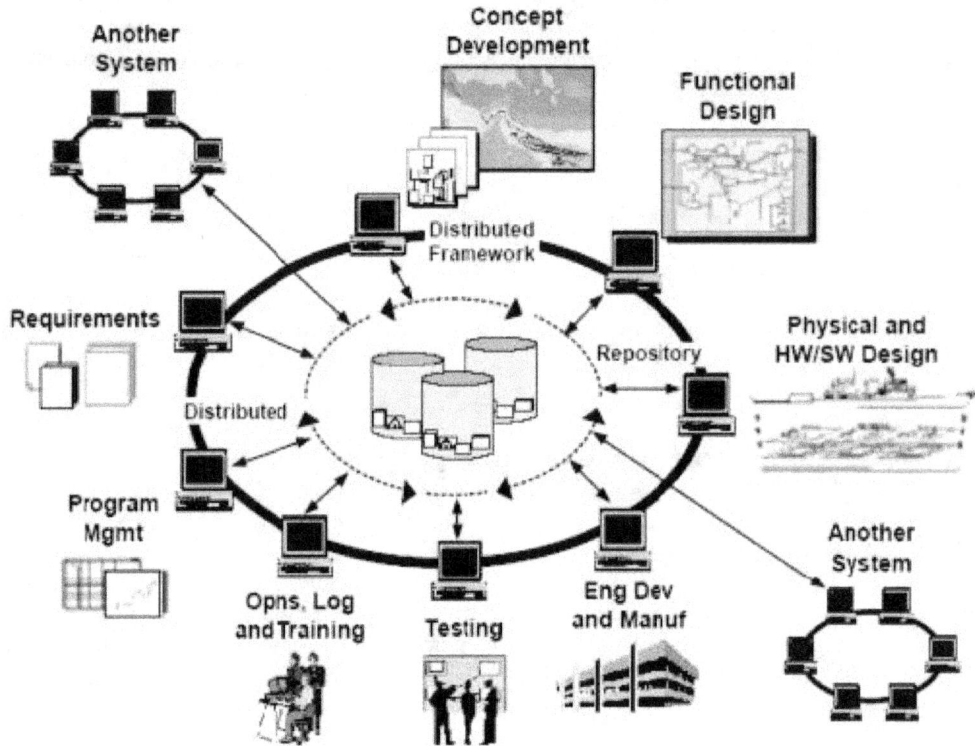

Example of the integrated use of Modelling and Simulation in Defence life cycle management. The modelling and simulation in this image is represented in the center of the image with the three containers.[10]

- Physical macrocosms

- Representational systems

- Syntactic models

- Emergent models

Model making in education, is an iterative exercise with students refining, developing and evaluating their models over time. This shifts learning from the rigidity and monotony of traditional curriculum to an exercise of students' creativity and curiosity. This approach utilizes the constructive strategy of social collaboration and learning scaffold theory. Model based learning includes cognitive reasoning skills where existing models can be improved upon by construction of newer models using the old models as a basis.[15]

"Model based learning entails determining target models and a learning pathway that provide realistic chances of under-standing." [16] Model making can also incorporate blended learning strategies by using web based tools and simulators, thereby allowing students to -

- Familiarize themselves with on-line or digital resources

- Create different models with various virtual materials at little or no cost

- Practice model making activity any time and any place

- Refine existing models

| Teacher assigns model-making assignment to Students in class or through online lesson tool (e.g. Blendspace) | Using physical material or online simulators (e.g. virtual lab), students create different models to understand the concept | Students - -study different models -record observations - conclude their understanding and discuss in classroom with teacher facilitating the learning process |

Flowchart Describing One Style of Model-based Learning

"A well-designed simulation simplifies a real world system while heightening awareness of the complexity of the system. Students can participate in the simplified system and learn how the real system operates without spending days, weeks or years it would take to undergo this experience in the real world." [17]

The teacher's role in the overall teaching and learning process is primarily that of a facilitator and arranger of the learning experience. He or she would assign the students, a model making activity for a particular concept and provide relevant information or support for the activity. For virtual model making activities, the teacher can also provide information on the usage of the digital tool and render troubleshooting support in case of glitches while using the same. The teacher can also arrange the group discussion activity between the students and provide the platform necessary for students to share their observations and knowledge extracted from the model making activity.

Model based learning evaluation could include the use of rubrics that assess the ingenuity and creativity of the student in the model construction and also the overall classroom participation of the student vis-a-vis the knowledge constructed through the activity.

It is important however to give due consideration to the following for successful model based learning to occur -

- Use of right tool at right time for a particular concept

- Provision within the educational setup for model making activity e.g. computer room with internet facility or software installed to access simulator or digital tool

18.5 See also

- Heuristic

- Grey box completion and validation

- Scientific visualization

- Simulation

- Statistical model
- Systems engineering
- Toy model

18.6   References

[1] Cartwright, Nancy. 1983. *How the Laws of Physics Lie*. Oxford University Press

[2] Hacking, Ian. 1983. *Representing and Intervening. Introductory Topics in the Philosophy of Natural Science*. Cambridge University Press

[3] Frigg and Hartmann (2009) state: "Philosophers are acknowledging the importance of models with increasing attention and are probing the assorted roles that models play in scientific practice". Source: Frigg, Roman and Hartmann, Stephan, "Models in Science", *The Stanford Encyclopedia of Philosophy* (Summer 2009 Edition), Edward N. Zalta (ed.), (source)

[4] Box, George E.P. & Draper, N.R. (1987). [Empirical Model-Building and Response Surfaces.] *Wiley*. p. 424

[5] Hagedorn, R. *et al.* (2005) http://www.ecmwf.int/staff/paco_doblas/abstr/tellus05_1.pdf *Tellus* 57A:219-233

[6] Leo Apostel (1961). "Formal study of models". In: *The Concept and the Role of the Model in Mathematics and Natural and Social*. Edited by Hans Freudenthal. Springer. p. 8-9 (Source)].

[7] Ritchey, T. (2012) Outline for a Morphology of Modelling Methods: Contribution to a General Theory of Modelling

[8] C. West Churchman, *The Systems Approach*, New York: Dell publishing, 1968, p.61

[9] Griffiths, E. C. (2010) What is a model?

[10] *Systems Engineering Fundamentals*. Defense Acquisition University Press, 2003.

[11] Pullan, Wendy (2000). *Structure*. Cambridge: Cambridge University Press. ISBN 0-521-78258-9.

[12] Fishwick PA. (1995). Simulation Model Design and Execution: Building Digital Worlds. Upper Saddle River, NJ: Prentice Hall.

[13] Sokolowski, J.A.,Banks, C.M.(2009). Principles of Modelling and Simulation. Hoboken, NJ: John Wiley and Sons.

[14] Lehrer, Richard; Schauble, Leona (2006). *The Cambridge Handbook of Learning Sciences*. Cambridge, UK: Cambridge University Press. p. 371. ISBN 978-0-521-84554-0.

[15] Nersessian, Nancy J (2002). *The Cognitive Basis of Science*. Cambridge, UK: Cambridge University Press. p. 133. ISBN 0-521-01177-9.

[16] Clement, JJ, Rea-Ramirez, Mary Anne (2008) *Model Based Learning and Instruction in Science* (2 ed.). Springer Science & Business Media. p. 45. ISBN 978-1-4020-6493-7.

[17] Blumschein, Patrick; Hung, Woei; Jonassen, David; Strobel, Johannes (2009). *Model-Based Approaches to Learning* (PDF). Netherlands: Sense Publishers. ISBN 978-90-8790-711-2.

18.7   Further reading

Nowadays there are some 40 magazines about scientific modelling which offer all kinds of international forums. Since the 1960s there is a strong growing amount of books and magazines about specific forms of scientific modelling. There is also a lot of discussion about scientific modelling in the philosophy-of-science literature. A selection:

- Rainer Hegselmann, Ulrich Müller and Klaus Troitzsch (eds.) (1996). *Modelling and Simulation in the Social Sciences from the Philosophy of Science Point of View. Theory and Decision Library*. Dordrecht: Kluwer.

- Paul Humphreys (2004). *Extending Ourselves: Computational Science, Empiricism, and Scientific Method*. Oxford: Oxford University Press.

- Johannes Lenhard, Günter Küppers and Terry Shinn (Eds.) (2006) "Simulation: Pragmatic Constructions of Reality", Springer Berlin.

- Tom Ritchey (2012). "Outline for a Morphology of Modelling Methods: Contribution to a General Theory of Modelling". In: *Acta Morphologica Generalis*, Vol 1. No 1. pp. 1-20.

- Fritz Rohrlich (1990). "Computer Simulations in the Physical Sciences". In: *Proceedings of the Philosophy of Science Association, Vol. 2*, edited by Arthur Fine et al., 507-518. East Lansing: The Philosophy of Science Association.

- Rainer Schnell (1990). "Computersimulation und Theoriebildung in den Sozialwissenschaften". In: *Kölner Zeitschrift für Soziologie und Sozialpsychologie* 1, 109-128.

- Sergio Sismondo and Snait Gissis (eds.) (1999). *Modeling and Simulation. Special Issue of Science in Context* 12.

- Eric Winsberg (2001). "Simulations, Models and Theories: Complex Physical Systems and their Representations". In: *Philosophy of Science* 68 (Proceedings): 442-454.

- Eric Winsberg (2003). "Simulated Experiments: Methodology for a Virtual World". In: *Philosophy of Science* 70: 105–125.

- Tomáš Helikar, Jim A Rogers (2009). "ChemChains: a platform for simulation and analysis of biochemical networks aimed to laboratory scientists". BioMed Central.

18.8   External links

- Models. Entry in the *Internet Encyclopedia of Philosophy*

- Models in Science. Entry in the *Stanford Encyclopedia of Philosophy*

- The World as a Process: Simulations in the Natural and Social Sciences, in: R. Hegselmann et al. (eds.), Modelling and Simulation in the Social Sciences from the Philosophy of Science Point of View, Theory and Decision Library. Dordrecht: Kluwer 1996, 77-100.

- Research in simulation and modelling of various physical systems

- Modelling Water Quality Information Center, U.S. Department of Agriculture

- Ecotoxicology & Models

- A Morphology of Modelling Methods. Acta Morphologica Generalis, Vol 1. No 1. pp. 1-20.

Template:Computer modelling

18.9 Text and image sources, contributors, and licenses

18.9.1 Text

2, Brandaray, Jowa fan, EmausBot, Orphan Wiki, Immunize, RA0808, RenamedUser01302013, Miss Manzana, Waidanian, Winner 42, Kierstend97, K6ka, Slawekb, ZéroBot, Fæ, Nomen4Omen, TheDinParis, Josve05a, Traxs7, East of Borschov, Ouzel Ring, Adcasaa, Chai Wei Jie, AvicAWB, Cobaltcigs, Wayne Slam, Cmathio, TyA, Piboy51, L Kensington, Alborzagros, Zayzya, Donner60, Zueignung, Orange Suede Sofa, ChuispastonBot, Jomomaindahouse, DASHBotAV, 28bot, SaschaWolff, ClueBot NG, Iiii I I I, Wcherowi, Dfarrell07, Braincricket, Widr, Helpful Pixie Bot, Langing, Technical 13, Koertefa, BG19bot, JoJaEpp, Kai Ojima, IiXbox Liive, John Cummings, Pieboy1012, MusikAnimal, Thumani Mabwe, Habitmelon, Wildog123, Quialls, Apurvgupta1996, Davethej, Death360, Brad7777, Trailspark, ColinKinloch, Markmathman, Burhan ud din Drabu, IkamusumeFan, Pratyush Sarkar, A1992, Dexbot, Aronmarshal, Auss00, Lugia2453, Frosty, Max Stardust, Nishantarya98, Coginsys, Ismail40, Ejkfnksdjf, JMCF125, Dddege, Akhilesh shende 15, Jigeesha choudary, Sportyort, Farai rabson, Marcuberte, Crow, Hommwock Smashr :)), Tim Topolski, Monkbot, Bub250, Kat+deven, JMP EAX, Barizo, Avengedsevenfoldjuggalo, Degenerate prodigy, Sumedh Tayade, KasparBot, Ctsevigny129 and Anonymous: 640

- **Product (mathematics)** *Source:* https://en.wikipedia.org/wiki/Product_(mathematics)?oldid=684640685 *Contributors:* AxelBoldt, The Epopt, Tarquin, XJaM, N8chz, Patrick, Michael Hardy, Wshun, Glenn, Mydogategodshat, Hollgor, Andrewman327, Hyacinth, Robbot, RedWolf, MathMartin, Tobias Bergemann, Giftlite, Herbee, Msm, Rich Farmbrough, MBisanz, RoyBoy, PAR, Bart133, Oleg Alexandrov, Linas, NeoUrfahraner, GlenPeterson, Terrx, Muchness, Ogai, Iancarter, Fs, Wknight94, 21655, Allens, SmackBot, Gilliam, Bluebot, Silly rabbit, Octahedron80, NYKevin, Lambiam, MvH, Bjankuloski06en-enwiki, IronGargoyle, EdC-enwiki, CzarB, CRGreathouse, Sopoforic, Karho.Yau, Thijs!bot, Escarbot, Cyclonenim, Bjoorn, Leuko, Yesitsapril, Maurice Carbonaro, SJP, Sunderland06, TheNewPhobia, VolkovBot, Philip Trueman, A4bot, Portalian, PaulColby, Happysailor, Paolo.dL, ClueBot, Dana boomer, XLinkBot, MystBot, Dsimic, SteveJothen, Addbot, Kongr43gpen, Skyezx, LaaknorBot, West.andrew.g, Numbo3-bot, PV=nRT, Zorrobot, Legobot, Yobot, Minvogt, KamikazeBot, SwisterTwister, AnomieBOT, IRP, AdjustShift, Materialscientist, Xqbot, Sionus, WaysToEscape, Erik9bot, Thehelpfulbot, FrescoBot, Mfwitten, HRoestBot, Sa'y, CobraBot, Jowa fan, EmausBot, Keilandreas, ZéroBot, Thine Antique Pen, BoredextraWorkvidid, ClueBot NG, Widr, Calabe1992, BG19bot, CityOfSilver, Cyclebear, BattyBot, IkamusumeFan, Dexbot, Brirush, Alisentas, Crow, WordSeventeen, Awesomewickedwhy and Anonymous: 95

- **Variable (mathematics)** *Source:* https://en.wikipedia.org/wiki/Variable_(mathematics)?oldid=685485093 *Contributors:* Michael Hardy, Rp, TakuyaMurata, Nickshanks, Robbot, Gandalf61, Tobias Bergemann, Giftlite, Micru, Macrakis, Kusunose, Iantresman, Mike Rosoft, Discospinster, Rgdboer, Kwamikagami, MattGiuca, Eclecticos, Silivrenion, Bgwhite, Phantomsteve, Reyk, True Pagan Warrior, SmackBot, RDBury, Georg-Johann, Rrburke, Cybercobra, Kashmiri, JForget, CRGreathouse, Myasuda, Gregbard, Cydebot, Thijs!bot, Marek69, Seaphoto, QuiteUnusual, Bongwarrior, P64, JamesBWatson, JaGa, R'n'B, Gill110951, Fylwind, Idioma-bot, VolkovBot, LokiClock, Philip Trueman, LimStift, Enviroboy, Symane, Gaelen S., SieBot, Steorra, Gerakibot, Flyer22, Michel421, Correogsk, Sean.hoyland, Melcombe, ClueBot, Mild Bill Hiccup, Excirial, He7d3r, Muhandes, SchreiberBike, Qwfp, Marc van Leeuwen, Mifter, Good Olfactory, Addbot, Some jerk on the Internet, Vchorozopoulos, Glane23, Tide rolls, Zorrobot, Luckas-bot, KamikazeBot, Reindra, AnomieBOT, Materialscientist, Holmes7893, The High Fin Sperm Whale, Xqbot, Jsharpminor, Isheden, Rodneidy, Erik9bot, Sławomir Biały, Boxplot, Pinethicket, Mrsmarenawalker, LittleWink, RedBot, TobeBot, LogAntiLog, Nataev, TjBot, DASHBot, EmausBot, Razertek, ZéroBot, Bollyjeff, Phrixussun, D.Lazard, Paulmiko, FrankFlanagan, BioPupil, Emperyan, ChuispastonBot, EdoBot, Txus.aparicio, ClueBot NG, Wcherowi, Satellizer, Cntras, Kevin Gorman, LightBringer, Widr, Jojo966, Ignacitum, HMSSolent, Wiki13, David815, AwamerT, Mark Arsten, Travelour, Thatemoverthere, GoShow, EuroCarGT, Dexbot, Lugia2453, Bulba2036, Jamesx12345, Nbeaver, YiFeiBot, Cubism44, Ashley angulo, Wikapedist, Thewikiguru1, Grayhawk22, Mhsh98, BrianPansky, Gmalaven, Rainboomcool, Gamingforfun365, Karissaisbae and Anonymous: 151

- **Coefficient** *Source:* https://en.wikipedia.org/wiki/Coefficient?oldid=680554136 *Contributors:* The Anome, Rade Kutil, Heron, Michael Hardy, Silverfish, Ffransoo, Charles Matthews, SchmuckyTheCat, Bkell, Hadal, Michael Snow, Tobias Bergemann, Marc Venot, Giftlite, Bovlb, Eequor, Mormegil, Discospinster, Paul August, Rgdboer, Sam Korn, Jumbuck, Msh210, Alansohn, Gene Nygaard, Crosbiesmith, WadeSimMiser, Magister Mathematicae, MarSch, Jameshfisher, Fresheneesz, TheGreyHats, Chobot, Roboto de Ajvol, YurikBot, RobotE, Pip2andahalf, Michael Slone, GeeJo, Shreshth91, S.L.-enwiki, Nucleusboy, Mad Max, DavidHouse-enwiki, Bota47, Haemo, Spliffy, SmackBot, Maksim-e-enwiki, Skizzik, IMacWin95, Octahedron80, Sidious1701, Cybercobra, Decltype, Amtiss, Cronholm144, Mets501, Igoldste, Hynca-Hooley, Iokseng, ST47, Biblbroks, UberScienceNerd, Epbr123, Braveorca, Escarbot, AntiVandalBot, Oddity-, Res2216firestar, JAnDbot, Bongwarrior, A Hauptfleisch, Granburguesa, JaGa, Hdt83, MartinBot, Tgeairn, Pharaoh of the Wizards, Lantonov, Enuja, Jarry1250, Signalhead, VolkovBot, Someguy1221, Broadbot, Maxim, Dogah, Xanstarchild, Paolo.dL, JackSchmidt, Denisarona, ClueBot, Deviator13, Gene93k, Uncle Milty, Niceguyedc, UKoch, DragonBot, Excirial, CrazyChemGuy, Estirabot, Thingg, DumZiBoT, Hotcrocodile, Marc van Leeuwen, Stickee, Gggh, CalumH93, Addbot, Proofreader77, Atethnekos, Fgnievinski, Ronhjones, Wikimichael22, Fluffernutter, AndersBot, Tide rolls, חיים רוזן, Zorrobot, Legobot, Luckas-bot, Yobot, SwisterTwister, Tempodivalse, Ciphers, Speller26, IRP, Piano non troppo, Darolew, Materialscientist, E2eamon, Capricorn42, Renaissancee, Mgtrevisan, 33rogers, LucienBOT, Bkerkanator, I dream of horses, Uknighter, Vrenator, علی ویکی, Diannaa, DARTH SIDIOUS 2, EmausBot, WikitanvirBot, RA0808, Darkfight, ZéroBot, John Cline, Chharvey, D.Lazard, ChuispastonBot, ClueBot NG, Gareth Griffith-Jones, Wcherowi, Helpful Pixie Bot, DBigXray, Mark Arsten, Peru Serv, Gunn1t, Omulae, GoShow, Makecat-bot, Lugia2453, Frosty, Cmckain14, Neitiznot, Tango303, Hollylilholly, Graceracer525, Carlos881, MinnieBeachBum1382 and Anonymous: 197

- **Ratio** *Source:* https://en.wikipedia.org/wiki/Ratio?oldid=683560166 *Contributors:* The Anome, Malcolm Farmer, LA2, Toby Bartels, Fubar Obfusco, Heron, Mei-enwiki, Michael Hardy, Kku, Dcljr, Ihcoyc, Samuelsen, Suisui, Glenn, Whkoh, Andres, Jiang, Charles Matthews, Sebastian Wallroth, Andrewman327, Furrykef, Hyacinth, David Shay, Omegatron, Bearcat, Robbot, Cyrius, Dave6, Psb777, Parasite, Giftlite, Tom harrison, Bovlb, IGEL, Kahkonen, Pgreenfinch, Sonett72, Alperen, Trevor MacInnis, Mike Rosoft, Discospinster, ElTyrant, Rich Farmbrough, YUL89YYZ, Foonly, Paul August, Fschoenm, Neko-chan, Phoenixdreaming, Mdf, Aude, Stesmo, Longhair, Sam Korn, Zachlipton, Alansohn, Bmh ca, Arthena, Ahruman, Wiccan Quagga, Velella, Vadakkan, Blaxthos, Oleg Alexandrov, Ampledata, Camw, Pol098, Acnetj, Mpatel, Btyner, MarcoTolo, Mandarax, Graham87, BD2412, Kbdank71, FreplySpang, Jshadias, Josh Parris, Jake Wartenberg, JHMM13, Jmcc150, The wub, Nigosh, Vuong Ngan Ha, RobertG, Gurch, TheDJ, AndriuZ, Chobot, The Rambling Man, Wavelength, Jimp, Pip2andahalf, Michael Slone, RadioFan, Stephenb, Yrithinnd, Gustavb, Rick Norwood, Holon, Dureo, Ruhrfisch, Dbfirs, Closedmouth, Pb30, Reyk, Heathhunnicutt, Smurrayinchester, Mikus, Cjfsyntropy, Allens, Cheesewizard, Fastifex, Theroachman, Amalthea, A bit iffy, SmackBot, RDBury, Cubs Fan, Jeppesn, KnowledgeOfSelf, Canthusus, BiT, Gilliam, Ohnoitsjamie, Skizzik, SchfiftyThree, Octahedron80, Nbarth, Baa, Baronnet, NYKevin, Berland, GVnayR, Flyguy649, Alca Isilon-enwiki, Eran of Arcadia, Sjock, Gobonobo, Bjankuloski06en-enwiki, IronGargoyle, 16@r, Mets501, Deflyer, Onionmon, Kvag, Esoltas, Nehrams2020, Iridescent, StephenBuxton, Chris53516, Blehfu, Tawkerbot2, AbsolutDan, ChrisCork, CRGreathouse, Filippos2, Asteriks, CBM, Dgw, MarsRover, Karenjc, Nilfanion, AndrewHowse, Zugvogel-enwiki, Joshua

Here. FlFvocals, AntiSpamBot, (jarbarf), Lbthrice, Bagpipeturtle, Anna Bailey, Richard D. LeCour, Cometstyles, King Toadsworth, Kvdveer, Xnuala, Moggie mn, Lights, My Core Competency is Competency, Cerberus0, Benjicharlton, CWii, Jeff G., Soliloquial, Philip Trueman, Showjumpersam, Anbellofe, GDonato, Anonymous Dissident, MackSalmon, Vgranucci, Itzcarlin, Carinemily, Insane-Contrast, NHRHS2010, Anameofmyveryown, Newbyguesses, Euryalus, Damorbel, ToXikyogHurt, Dueeex2, Keilana, Danielgrad, Green-eyed girl, Reinderien, Guy-calledryan, Virginia malone, Oxymoron83, Nk.sheridan, Phil Lu, Sunrise, Melcombe, Lloydpick, Emptymountains, ClueBot, Artichoker, Alpha Beta Epsilon, Lawrence Cohen, WDavis1911, Mild Bill Hiccup, Polyamorph, Robert Skyhawk, Abrech, Autoplayer, Tnxman307, Lenary, La Pianista, Calor, Aitias, Nebula2357, XLinkBot, Marc van Leeuwen, Rror, Protectthehuman, Eleven even, Tayste, Samidril, Addbot, Jester-hasissues, Wheelman200, Fgnievinski, CanadianLinuxUser, TundraGreen, Tedtoal, Luckas-bot, 2D, The Grumpy Hacker, Grebaldar, THEN WHO WAS PHONE?, AnomieBOT, Southernwolfie, PowerUserPCDude, ArthurBot, GenQuest, Br77rino, ProtectionTaggingBot, Thosjleep, Banak, Jamsyst, Redrose64, Machn, Ver-bot, Duoduoduo, ZéroBot, JA(000)Davidson, Dreispt, ClueBot NG, Shovan Luessi, Helpful Pixie Bot, BG19bot, Vadering, ConradMayhew, Illia Connell, JYBot, Mervat Salman, Brirush, Yardimsever, Rangeblock victim, ElHef, EoRdE6 and Anonymous: 511

- **Cartesian coordinate system** *Source:* https://en.wikipedia.org/wiki/Cartesian_coordinate_system?oldid=673154275 *Contributors:* Damian Yerrick, Chuck Smith, Bryan Derksen, Zundark, Tarquin, Mark Ryan, Andre Engels, Heron, Montrealais, Patrick, Michael Hardy, Deljr, JWSchmidt, Александр, AugPi, Skyfaller, Smack, Pizza Puzzle, Nikola Smolenski, Drz-enwiki, Emperorbma, Charles Matthews, Timwi, Dysprosia, Colipon, The Anomebot, Maximus Rex, Bevo, Chuunen Baka, Robbot, Romanm, Modulatum, Sverdrup, Henrygb, Prara, Michael Snow, Jleedev, Enochlau, Snobot, Giftlite, DocWatson42, Jason Quinn, Jorge Stolfi, Manuel Anastácio, SoWhy, MrMambo, Joseph Myers, C4-enwiki, Adashiel, Discospinster, Rich Farmbrough, Guanabot, Paul August, Bender235, Elwikipedista-enwiki, BenjBot, Edwinstearns, Rgdboer, Mavhe, Che090572, Kjkolb, Larry V, HasharBot-enwiki, OGoncho, Jumbuck, Alansohn, SnowFire, PAR, Yossiea-enwiki, Fordan, Dionoea, Kbolino, Falcorian, Oleg Alexandrov, Mel Etitis, Woohookitty, Linas, LOL, Firefishy, Mpatel, Waldir, Ruziklan, Palica, Tony1849, Graham87, Jobnikon, BD2412, Rjwilmsi, MJSkia1, Ygrek, OneWeirdDude, MarSch, Jmcc150, Salix alba, Brighterorange, Titoxd, FlaBot, Mathbot, Nihiltres, Maxal, Cmbrannon, Slant, Fresheneesz, Srleffler, Chobot, Helios, DVdm, YurikBot, Wavelength, RobotE, Charles Gaudette, Pip2andahalf, RussBot, Hede2000, Piet Delport, Gustavb, Byj2000, RabidDeity, Dhollm, Bucketsofg, Dbfirs, Cheeser1, DeadEyeArrow, Dast, MathsIsFun, HereToHelp, Tony Liao-enwiki, JLaTondre, Gesslein, Archer7, Allens, Katieh5584, Meegs, Nekura, DVD R W, Sardanaphalus, SmackBot, RDBury, Adam majewski, Incnis Mrsi, Delldot, Canthusus, ParlorGames, Skizzik, Mirokado, Kurykh, Miquonranger03, Silly rabbit, Basalisk, Octahedron80, Nbarth, DHN-bot-enwiki, Hongooi, Antonrojo, Can't sleep, clown will eat me, RedHillian, SundarBot, Cameron Nedland, Memming, Downtown dan seattle, Doodle77, Andeggs, Sadi Carnot, Andrei Stroe, Lambiam, Dbtfz, Kusarbo, Cronholm144, Bjankuloski06en-enwiki, Aleenf1, IronGargoyle, 041744, JHunterJ, Stwalkerster, Mets501, Wizard191, Maelor, JForget, CmdrObot, Jackzhp, Dgw, NickW557, 345Kai, WeggeBot, Yaris678, WillowW, LouisBB, ST47, Edgerck, Juansempere, DumbBOT, TJ09, Vanished User jdksfajlasd, Zalgo, Lanepierce, Thijs!bot, Epbr123, Headbomb, Federhalter, Escarbot, Cyclonenim, AntiVandalBot, Gioto, Mhaitham.shammaa, Modernist, JAnDbot, Chausx, PhilKnight, DeclinedShadow, Magioladitis, Connormah, VoABot II, Catslash, Mrfence, CattleGirl, Fabrictramp, Srice13, Maniwar, David Eppstein, DerHexer, MartinBot, Rettetast, Snozzer, J.delanoy, Captain panda, Sasajid, Trusilver, Numbo3, Century0, Cpiral, Lantonov, Stan J Klimas, Samtheboy, NewEnglandYankee, Davecrosby uk, CardinalDan, RJASE1, Idioma-bot, KillerOfThem, VolkovBot, JohnBlackburne, Mun206, TXiKiBoT, Anonymous Dissident, Steven J. Anderson, Martin451, Uncaringgunner, Domitius, Topherjasmin09, Andy Dingley, Dirkbb, Blindman.rms, AlleborgoBot, Logan, Katzmik, SieBot, Euryalus, Yintan, Joaosampaio, Flyer22, Man It's So Loud In Here, Paolo.dL, Oxymoron83, Atmamatma, Jurlinga, Hello71, Hobartimus, Svick, Anchor Link Bot, Randomblue, Escape Orbit, Martarius, ClueBot, The Thing That Should Not Be, Mild Bill Hiccup, DragonBot, Excirial, Abrech, SockPuppetForTomruen, Thingg, Mattreedywiki, Johnuniq, Darkicebot, CaptainVideo890, Skunkboy74, Vanostran, Hyperweb79, Addbot, Binary TSO, DougsTech, Fgnievinski, Blethering Scot, Jncraton, Fieldday-sunday, Leszek Jańczuk, MrOllie, Allliam, TheFreeloader, Tide rolls, Alanfeynman, Lightbot, Nobono9, Zorrobot, Ⴑⴓⴟⴟⴞⴈ, Wmplayer, Wwannsda, Luckas-bot, Yobot, Fraggle81, Anypodetos, Nallimbot, Vltava 68, Tempodivalse, AnomieBOT, 1exec1, Rajmathi mehta, Piano non troppo, Ulric1313, Materialscientist, Citation bot, Oftopladb, Xqbot, Waffleman12, Jeffwang, NorbDigiBeaver, Almabot, Frosted14, Red van man, A.amitkumar, FrescoBot, Appropo, Majopius, Masterknighted, Rhino bucket, Wireless Keyboard, Þjóðólfr, Pmokeefe, Jsjunkie, Rogiemac, Shanmugamp7, Rausch, كاشف عقيل, Lotje, Colin Cochrane, Dasteve, Suffusion of Yellow, Shanker Pur, NameIsRon, Timh3221, Slon02, EmausBot, Acather96, WikitanvirBot, GoingBatty, RA0808, Wikipelli, Slawekb, CanonLawJunkie, Knight1993, Junelvillejo, Quondum, MonoAV, Chewings72, WMC, ClueBot NG, Gareth Griffith-Jones, KlappCK, Wcherowi, MelbourneStar, Satellizer, Movses-bot, Widr, Helpful Pixie Bot, DBigXray, Popsh, Papadim.G, Questionefisica, Mark Arsten, Blue Mist 1, Phl.jns, Nbrothers, Pratyya Ghosh, SergeantHippyZombie, Radio15dude, ChrisGualtieri, Bigloser12345loser, EuroCarGT, Ramesepirate, Ducknish, Kelvinsong, Kingbowen, Webclient101, Indiana State, RazrRekr201, Jc86035, Acetotyce, ProtossPylon, Lickturtles, SamX, Wamiq, Sicaeffect, Ginsuloft, Primalshell, JAaron95, Stamptrader, Andreatristan, Akhilburle, Wilson Widyadhana, Roshmita, Elsa1098, Hdkeudhdjisjedu and Anonymous: 474

- **Hyperbolic coordinates** *Source:* https://en.wikipedia.org/wiki/Hyperbolic_coordinates?oldid=676422304 *Contributors:* Kwertii, Charles Matthews, Jason Quinn, Edudobay, Rgdboer, Longhair, Ron Ritzman, Linas, Grammarbot, RussBot, Crasshopper, SmackBot, RDBury, Lambiam, CBM, Myasuda, Christian75, Rocchini, Lantonov, AlnoktaBOT, Thurth, Cuddlyable3, Mleconte, Qwfp, Addbot, Tide rolls, Ciphers, LilHelpa, Lu-cienBOT, Helpful Pixie Bot, SandraShklyaeva, Mark viking, Lemnaminor, Loraof and Anonymous: 11

- **Hyperbola** *Source:* https://en.wikipedia.org/wiki/Hyperbola?oldid=681724600 *Contributors:* AxelBoldt, Brion VIBBER, Tarquin, AstroNom Xaonon, Arvindn, PierreAbbat, Patrick, Michael Hardy, Tongpoo, Wapcaplet, Cyde, Ellywa, Ojs, AugPi, Hemmer, Charles Matthews, Dysprosia, Saltine, Alembert-enwiki, Sabbut, Tjdw, Donarreiskoffer, Fredrik, CHz, Wereon, Jleedev, Tosha, Decrypt3, Giftlite, Jyril, Gene Ward Smith, Harp, Wwoods, Alison, Tom-, Jorge Stolfi, Macrakis, LiDaobing, Alberto da Calvairate-enwiki, Melikamp, Jossi, Secfan, Maximaximax, Sam Hocevar, Didactohedron, Qef, Archer3, JTN, Discospinster, FiP, Paul August, Brian0918, Rgdboer, Spoon!, Viriditas, Reubot, Alansohn, Mickeyreiss, Wtmitchell, Jheald, Ceyockey, Deror avi, Postrach, Velho, Woohookitty, Linas, Thruston, SeventyThree, Graham87, Magister Mathematicae, Rjwilmsi, HannsEwald, Zhurovai, FlaBot, Mathbot, Alphachimp, Chobot, Krishnavedala, DVdm, Twoeyedhuman, Stephen Compall, YurikBot, Wavelength, Vagodin, Sceptre, Akamad, Rick Norwood, Grafen, Raven4x4x, Cheeser1, BOT-Superzerocool, Cmglee, SmackBot, RDBury, PEHowland, Reedy, Jeppesn, David.Mestel, Unyoyega, AndyZ, Eskimbot, Hmains, Keegan, MalafayaBot, Dabigkid, DHN-bot-enwiki, Colonies Chris, Hengsheng120, Matthijs, SundarBot, RandomP, Acdx, Spiritia, Phancy Physicist, MarkSutton, Mets501, Laurens-af, Tawkerbot2, The Letter J, Vaughan Pratt, Mikiemike, CBM, Kewldude606, Frankshiv, Nap, Doctormatt, WillowW, MuKinpatsuDijou, Goldencako, Thijs!bot, Wikid77, RatedRestricted, A3RO, Geekdog, Jj137, Phanerozoic, Eleos, JAnDbot, Greensburger, LittleOldMe, Ibapah, Maniwar, Beherkas, Donnyton, David Eppstein, Styrofoam1994, MartinBot, Geranston, Captain panda, SharkD,

Krishnachandranvn, Policron, Zojj, 2help, Bcnof, LaughingVulcan, 106er, Gogobera, Deor, VolkovBot, Joeoettinger, AlnoktaBOT, TXiKi-BoT, Anonymous Dissident, Melsaran, LeaveSleaves, Inductiveload, Deipnosopher, Jesin, Eubulides, Seresin, SieBot, Unamed102, Ceroklis, Gerakibot, MyNameIsHi, Wombatcat, Dhexus–enwiki, Nic bor, Sgtzac1, ClueBot, Plastikspork, Arakunem, AlasdairGreen27, Blanchardb, LizardJr8, DragonBot, Tylerdmace, Cenarium, Qwfp, Deutschnummereins, Crowsnest, DumZiBoT, Noeatingallowed, Addbot, Fgnievin-ski, Computer Guy 990, Ronhjones, MrOllie, AnnaFrance, Kiril Simeonovski, Zorrobot, Luckas-bot, Yobot, Stameose, THEN WHO WAS PHONE?, AnomieBOT, Götz, To Fight a Vandal, Pontificalibus, RibotBOT, Rainald62, Constructive editor, FrescoBot, GuidoB, Pinethicket, Bookerj, RedBot, 1to0to-1, Jujutacular, Turian, Rausch, Reconsider the static, December21st2012Freak, DixonDBot, Michael9422, Duoduo-duo, Garglax, EmausBot, Mz7, Tim Zukas, Puffin, 28bot, Anita5192, ClueBot NG, Plantkeeper, Wcherowi, Chester Markel, Frietjes, Joel B. Lewis, Anxiousswift, Hallows AG, Davidiad, Six55eee6, Wing139, Benlcox, Hmainsbot1, Argon34, Ag2gaeh, Bigsexy58, Trixie05, Library Guy, Joshwitges, Loraof, KasparBot, Kafishabbir and Anonymous: 217

- **Hyperbolic growth** *Source:* https://en.wikipedia.org/wiki/Hyperbolic_growth?oldid=628019096 *Contributors:* The Anome, Charles Matthews, Giftlite, Gyrofrog, Bender235, Nigelj, Army1987, Alansohn, Sligocki, Diza, Dbfirs, SmackBot, Portillo, Nbarth, CRGreathouse, CmdrObot, Soimless, Phanerozoic, Athkalani–enwiki, Ryan032, Mark v1.0, Ctxppc, Athkalani2000, 1ForTheMoney, Editor2020, Tayste, Addbot, Yobot, Legobot II, Br77rino, GrouchoBot, David815, Fylbecatulous and Anonymous: 18

- **Multiplicative inverse** *Source:*https://en.wikipedia.org/wiki/Multiplicative_inverse?oldid=682637979*Contributors:*Toby Bartels, Patrick, Michael Hardy, Ixfd64, Eric119, Glenn, Timwi, Dcoetzee, Frazzydee, Robbot, Giftlite, Fropuff, No Guru, Pmanderson, Trevor MacInnis, Chrisjwmartin, Discospinster, Rich Farmbrough, ArnoldReinhold, Notthepainter, EmilJ, Bobo192, Wood Thrush, Haham hanuka, Jumbuck, Alansohn, Tobych, Jheald, Natalya, Georgia guy, Ruud Koot, WadeSimMiser, MFH, Marudubshinki, Dysepsion, Jshadias, Josh Parris, Math-bot, Chobot, Jersey Devil, Algebraist, Pandelon, RussBot, Michael Slone, KSmrq, Gotfrie, Zwobot, Bota47, Wknight94, Square87–enwiki, Arthur Rubin, JLaTondre, Gesslein, Leon2323, Ghazer–enwiki, Bo Jacoby, KocjoBot–enwiki, GraemeMcRae, Dan Hoey, Octahedron80, HoodedMan, UU, Wen D House, Dreadstar, Tbjw, Rigadoun, Mgiganteus1, Jim.belk, Slakr, Mets501, Avant Guard, Levineps, Madmath789, CRGreathouse, CBM, OMGsplosion, Doctormatt, He Who Is, Omicronpersei8, Thijs!bot, Epbr123, FreeKresge, AntiVandalBot, Mhaitham.shammaa, Salgueiro–enwiki, TuvicBot, JAnDbot, Greensburger, Lawilkin, VoABot II, Toomai Glittershine, MyNameIsNeo, David Eppstein, Rettetast, Anaxial, Leyo, Zorakoid, Uncle Dick, Mike.lifeguard, WarthogDemon, OohBunnies!, Montchav, R00723r0, Philip Trueman, TXiKiBoT, Nxavar, Gauge00, Kmhkmh, Dmcq, Logan, Ben Boldt, Dogah, Caltas, Jackpots, Smoby10, Oxymoron83, JackSchmidt, ClueBot, Justin WSmith, Cb4astros, BarretB, Marc van Leeuwen, Rror, Coolbeans39, Stephen Poppitt, Addbot, Ramu50, Friginator, Ronhjones, Glane23, Numbo3-bot, Tide rolls, Narnaja, Snaily, Luckas-bot, Yobot, PMLawrence, Piano non troppo, Kingpin13, Moipaulochon, DirlBot, Xqbot, RibotBOT, Mothemessfather, Robo37, Hoo man, RedBot, Duoduoduo, Copistopplayer, WikitanvirBot, ZéroBot, Quondum, Vanished userfijtji34toksdcknqrja54yoimascj, NTox, Lyleq, ClueBot NG, HMSSolent, Calabe1992, AvocatoBot, Hillcrest98, Yomama719, Everymorning, Lordofbartonpark, Ginsuloft, Loraof and Anonymous: 167

- **Linear function** *Source:* https://en.wikipedia.org/wiki/Linear_function?oldid=681005892 *Contributors:* Shd–enwiki, XJaM, Olivier, Patrick, Michael Hardy, JakeVortex, Kku, TakuyaMurata, Drz–enwiki, Stan Lioubomoudrov, Saltine, Tobias Bergemann, Alan Liefting, Giftlite, Foo-bar, Almit39, Imjustmatthew, Vsmith, HCA, Paul August, El C, Shanes, Kappa, Alansohn, Oleg Alexandrov, Ian Pitchford, PlatypeanArch-cow, Margosbot–enwiki, RexNL, Fresheneesz, Forzaferrara, Krishnavedala, DVdm, YurikBot, RobotE, RussBot, KSmrq, Terfili, Panscient, Thiim–enwiki, 21655, Gesslein, Nojhan, Luk, Schizobullet, SmackBot, Adam majewski, Slashme, Gilliam, Skizzik, Bluebot, VMS Mosaic, Cybercobra, J. Finkelstein, 16@r, Dicklyon, Stalker314314, Beve, Wafulz, Marcusyoder, CBM, Myasuda, Zginder, Marek69, AntiVandalBot, JAnDbot, Leuko, Magioladitis, MartinBot, Zeus, Dorvaq, Qrystal, J.delanoy, Rhinestone K, McSly, Gombang, Celtic Minstrel, LokiClock, Thurth, Orie0505, Geometry guy, Dsignoff, Tugbug, SieBot, K. Annoyomous, Jauerback, Nathanvermeulen, Darkmyst932, ClueBot, Fyyer, The Thing That Should Not Be, Smithpith, Alksentrs, Zack wadghiri, CrazyChemGuy, Darren23, 7, SoxBot III, Addbot, Fyrael, Fgnievinski, SpBot, Ozob, PV=nRT, Zorrobot, Jarble, Luckas-bot, TaBOT-zerem, Daniel 1992, AnomieBOT, Erel Segal, Kingpin13, Xqbot, RibotBOT, Ssupafly, Dougofborg, Pinethicket, Tcnuk, Extra999, Duoduoduo, Igogo3000, EmausBot, RA0808, ZéroBot, Myah07, Quondum, D.Lazard, Makecat, L Kensington, Bill william compton, ClueBot NG, Wcherowi, This lousy T-shirt, Lanthanum-138, Marechal Ney, Widr, Hagoth, MusikAnimal, J991, Solomon7968, Fuhaoming, 铁铁的火大, Minsbot, Lugia2453, Brirush, Mark viking, Shivd18, Ginsuloft, Bilorv, Derektuttle and Anonymous: 163

- **Exponential function** *Source:* https://en.wikipedia.org/wiki/Exponential_function?oldid=682819966 *Contributors:* AxelBoldt, Bryan Derk-sen, Tarquin, Ap, XJaM, Christian List, Enchanter, Heron, B4hand, Patrick, Michael Hardy, Wshun, MartinHarper, Ixfd64, Dcljr, TakuyaMu-rata, GTBacchus, Eric119, Minesweeper, ArnoLagrange, Ellywa, Cyp, Stevenj, Kingturtle, Poor Yorick, Smack, Revolver, Charles Matthews, Timwi, Dysprosia, AndrewKepert, Omegatron, Wizow–enwiki, Kwantus, Robbot, Fredrik, Mattblack82, Gandalf61, MathMartin, Sverdrup, Jleedev, Tobias Bergemann, Giftlite, MathKnight, Ssd, Sietse, Nayuki, Cam, Demon1hung, AHM, Ricky–enwiki, Kusunose, Tsemii, Plutor, Brianjd, Guanabot, ObsessiveMathsFreak, ArnoldReinhold, Eric Shalov, Ascánder, Sam Derbyshire, MuDavid, Paul August, Zenohockey, RoyBoy, Vdm, Army1987, Cmdrjameson, ::Ajvol::, Haham hanuka, Spitzl, Sligocki, Pjacklam, PAR, Caesura, Oleg Alexandrov, Abanima, Waabu, Prashu, Linas, Shreevatsa, Isnow, Gerbrant, Graham87, Chenxlee, Josh Parris, Rjwilmsi, Isaac Rabinovitch, NeonMerlin, AndyKali, Ohanian, MikeJ9919, Bmicomp, Glenn L, Chobot, Jersey Devil, YurikBot, Wavelength, Piet Delport, Stephenb, Rsrikanth05, Amcfreely, Db-firs, Ninly, Petri Krohn, Flowersofnight, Gesslein, MagneticFlux, Bo Jacoby, Jinxs, SmackBot, Moeron, PizzaMargherita, Eskimbot, Alsandro, Emj, Rgrizza, Kurykh, Dustimagic, Octahedron80, Zven, Ewjw, Mhym, Memming, Henning Makholm, Richard0612, Andrei Stroe, Sasha-toBot, Lambiam, LinuxDude, Bilboq, Daphne A, Stalker314314, TastyPoutine, BranStark, A. Pichler, Rdunn, CRGreathouse, HenningTh-ielemann, Gogo Dodo, He Who Is, DumbBOT, Thijs!bot, Epbr123, Ucanlookitup, Dugwiki, Albert1ls, BigJohnHenry, Myanw, JAnDbot, Magioladitis, Tt 225, David Eppstein, JaGa, MartinBot, J.delanoy, Pharaoh of the Wizards, Maurice Carbonaro, Nigholith, Ryan Postleth-waite, Chriswiki, Policron, 2help, Dessources, DavidCBryant, Secleinteer, Salte45, Idioma-bot, Sheliak, VolkovBot, Jeff G., JohnBlackburne, Philip Trueman, TXiKiBoT, Anonymous Dissident, AJRobbins, Dmcq, Kbrose, Brocksimpson, YohanN7, SieBot, Tiddly Tom, Markedis-onehua, Cole SWE, JCLately, Thehotelambush, Milo257, Witepa, Anchor Link Bot, Mygerardromance, Oekaki, ClueBot, The Thing That Should Not Be, Smithpith, BigMike718, Thegeneralguy, Fasttimes68, Jsondow, Ottawa4ever, Unenough, JDPhD, Ultra.Power, Pitt SATSA, Gerhardvalentin, Kal-El-Bot, Calcio1, Fgnievinski, WardenWalk, Cst17, Favonian, TStein, Ozob, Ehrenkater, ‎אדם-סריג, Legobot, Yobot, Tohd8BohaithuGh1, LGB, QueenCake, Swister-Twister, IW.HG, AnomieBOT, Götz, Maarwaan, Citation bot, Akilaa, Wiggin Tree, Fact-Spewer, Bdmy, BISCUITKID, Tyrol5, Isheden, Inferno, Lord of Penguins, Omnipaedista, Oscarjquintana, SassoBot, Gordonrox24, GESICC, Adler.fa, Roland Crosby, HJ Mitchell, Pratik.mallya, Petsam007, Shuroo, Maggyero, Pinethicket, Oveckin 08, Rushbugled13, Yahia.barie,

Meaghan, Math.geek3.1415926, 777sms, Περίεργος, Threefourfive345, Tbhotch, MMS2013, Tom38-enwiki, DoRD, EmausBot, GeneralCheese, Tommy2010, Wtuvell, Thecheesykid, Almacantar, Gribeauval, ZéroBot, Josve05a, Chharvey, Quondum, Loc041294, Sigma0 1, Ashi1564, ClueBot NG, Accelerometer, Satellizer, Griffbo, Helpful Pixie Bot, Wackywill1001, Jan Spousta, Mark Arsten, Brad7777, IkamusumeFan, ChrisGualtieri, Sankar1408, JYBot, Webclient101, Vbaculum, Someonewhoisme, HelicopterLlama, Mark viking, Fycafterpro, ValuableAppendage, Felixabbud, Shannonbecca, Peiffers, Raycheng200, Elizafish, MathoMathew, KasparBot, De la Marck, Owen Dostie, Nuwanst and Anonymous: 273

- **Logarithm** *Source:* https://en.wikipedia.org/wiki/Logarithm?oldid=685032767 *Contributors:* AxelBoldt, Peter Winnberg, Brion VIBBER, Mav, Rjstott, Arvindn, Bth, Stevertigo, Edward, Bdesham, Patrick, Michael Hardy, GABaker, Zocky, Wshun, TakuyaMurata, GTBacchus, Loisel, Eric119, Iluvcapra, Minesweeper, Egil, Looxix-enwiki, Ellywa, Ahoerstemeier, Cyp, Stan Shebs, 1jon, Mark Foskey, Julesd, Lupinoid, AugPi, Poor Yorick, GRAHAMUK, Dwo, Artoo-enwiki, Charles Matthews, Dcoetzee, Dysprosia, Jitse Niesen, Doradus, WhisperToMe, Tpbradbury, Hyacinth, Saltine, Itai, AndrewKepert, Omegatron, Wazow-enwiki, Jonhays0, Robbot, Fredrik, Fifelfoo, Modulatum, Gandalf61, Henrygb, Gidonb, Robinh, Ruakh, HaeB, Dmn, Jleedev, Tobias Bergemann, Enochlau, Decrypt3, Centrx, Giftlite, Gene Ward Smith, Luis Dantas, BenFrantzDale, Tom harrison, Lupin, Herbee, Peruvianllama, Bensaccount, Niteowlneils, Guanaco, Rjvanco, Nova77, CryptoDerk, Noe, MarkSweep, Piotrus, Kaldari, MacGyverMagic, Fuper, Gauss, Mike Storm, Elektron, Pmanderson, Icairns, Joyous!, Eliazar, Grunt, ELApro, Thorwald, Brianjd, Mormegil, Discospinster, Rich Farmbrough, Rhobite, ObsessiveMathsFreak, Pie4all88, ArnoldReinhold, Will2k, 1pezguy, MuDavid, Harriv, Paul August, Night Gyr, Compie, Billion, Pmetzger, RJHall, Pt, MBisanz, El C, Rgdboer, Shanes, Rimshot, Wareh, Andrewpmack-enwiki, Touriste, Army1987, Harley peters, Reinyday, Nk, Mjager, Ardric47, Obradovic Goran, Hesperian, Wrs1864, Hagerman, Ultra megatron, Alansohn, Arthena, Rxc, Wiki-uk, AzaToth, Bz2, Sligocki, PAR, Pion, Snowolf, KingTT, Count Iblis, Dirac1933, H2g2bob, Woodstone, Kusma, Gene Nygaard, MIT Trekkie, HenryLi, Forderud, Oleg Alexandrov, Stephen, Waabu, Corsairstw, OwenX, Linas, LOL, Robert K S, WadeSimMiser, Matijap, Hdante, Mpatel, Eleassar777, Bbatsell, Prashanthns, Reddwarf2956, CannibalSmith, Leapfrog314, Dysepsion, Merideth, Graham87, Magister Mathematicae, Jetekus, David Levy, Yurik, Jclemens, Rjwilmsi, JVz, Koavf, Salix alba, HappyCamper, NeonMerlin, Bubba73, Docether, Ohanian, Bfigura, Ucucha, Sango123, Yamamoto Ichiro, Titoxd, StuartBrady, Alejo2083, FlaBot, VKokielov, RobertG, SeptimusOrcinus, Musical Linguist, Mathbot, MacRusgail, Nivix, RexNL, Gurch, Valermos, Ichudov, AntonioDsouza, Fresheneesz, Kri, Glenn L, Chobot, Krishnavedala, Zath42, DVdm, Twoeyedhuman, Bgwhite, Hall Monitor, Krysith, YurikBot, Wavelength, Angus Lepper, Jimp, Wolfmankurd, Vecter, Agent Foxtrot, Sarranduin, Taejo, KSmrq, Mpfrank, Yyy, NawlinWiki, Wiki alf, Grafen, Petter Strandmark, Trovatore, Luks, Vaclav Haisman, Ja'Achan, Gadget850, Wknight94, PGPirate, Zunaid, Ninly, Closedmouth, Ajsdecepida, Netrapt, JahJah, Cjfsyntropy, Gesslein, Shastra, Maxsupereme, Katieh5584, Aaron Will, Teply, GrinBot-enwiki, Cmglee, Sbyrnes321, Attilios, SmackBot, RDBury, Promsan, Khfan93, Incnis Mrsi, Melchoir, David Shear, Jagged 85, Eskimbot, Rajah9, BiT, Bromskloss, Durova, IloveMP2yea, Stubblyhead, B00P, Janm67, Rafterman-enwiki, Octahedron80, Nbarth, Kostmo, Southcaltree, Can't sleep, clown will eat me, Chlewbot, Vanished User 0001, Snowmanradio, Georg-Johann, Rrburke, GRuban, Geoboe84, Whpq, Brutha-enwiki, Grover cleveland, Epachamo, Savidan, RandomP, DMacks, Adrianiu, Dnavarro, Charivari, SashatoBot, Lambiam, TheHorse'sMouth, Chrisandtaund, Mbeychok, Brian Gunderson, Olin, NongBot-enwiki, IronGargoyle, 16@r, A. Parrot, Paradoxsociety, Noah Salzman, Owlbuster, Dicklyon, Optakeover, CUTKD, Eridani, Mets501, Pezant, Dr.K., MTSbot-enwiki, Kvng, Mfield, KyleP, Asyndeton, MystRivenExile, Iridescent, Joseph Solis in Australia, J Di, Shoshonna, Tawkerbot2, JRSpriggs, Cryptic C62, Nutster, JForget, Trombe29, CRGreathouse, Makeemlighter, CBM, 345Kai, CKozeluh, HenningThielemann, Shanoman, Yaris678, Doctormatt, Meznaric, Cantras, WillowW, Mike Christie, Dominicanpapi82, Carifio24, Julian Mendez, He Who Is, Oz an, The Jimmy, Biggoggs, Biblbroks, ThomasGHenry, Brad101, Kansas Sam, MayaSimFan, Mitchoyoshitaka, Ziggot, Malleus Fatuorum, Thijs!bot, Chu333222, KCliffer, Headbomb, Najro, Cardboard Moose, Jojan, RobHar, Tocharianne, Link hyrule5, Northumbrian, Hmrox, Jkwilson, AntiVandalBot, WinBot, Courtjester555, Prolog, RapidR, Jj137, Zachwoo, Lfstevens, Schwilgue, JAnDbot, MER-C, Bekant, Ricardo sandoval, Andonic, Drvinginshlagin, .anacondabot, Magioladitis, WolfmanSF, Bongwarrior, VoABot II, Ishikawa Minoru, Think outside the box, Jakob.scholbach, Sikory, Crunchy Numbers, Adrian J. Hunter, Darkrync183, Allstarecho, David Eppstein, Error792, Glen, Rajpaj, DerHexer, Khalid Mahmood, DAVIDY, Squidonius, Tercer, Monurkar-enwiki, Gwern, MartinBot, Franp9am, Racepacket, Fuzziqersoftware, NAHID, Retarius, Glrx, Gah4, Obscurans, PhageRules1, J.delanoy, Shining Arcanine, DrKay, EscapingLife, Inimino, Jacques-laporte, SharkD, Algebragirl, ShamusOmalley, NewEnglandYankee, Nwbeeson, SmilesALot, Policron, Bohianite, Wng z3r0, 2help, DavidCBryant, Tiggerjay, Jamesontai, DeFaultRyan, Rawr rawr roar, Jarry1250, Salte45, Idioma-bot, JonShops, Nitroshockwave, JohnBlackburne, Orthologist, LokiClock, Yiming689, Lexein, Gaianauta, Soliloquial, PMajer, Philip Trueman, TXiKiBoT, GimmeBot, Leo628, GDonato, Lechatjaune, Anonymous Dissident, Qxz, Someguy1221, Pakalomattam, Jave pogi, PaulTanenbaum, LeaveSleaves, PFrisbie, Geometry guy, Zahical, Wolfrock, Synthebot, Falcon8765, @pple, AJRobbins, WatermelonPotion, Dmcq, Symane, Scottywong, Nschoem, SieBot, Tresiden, Impolius Impus, Graham Beards, Scarian, Gerakibot, SE16, Cwkmail, GeiwTeol, Cb77305, Bxn1358, Garde, Cffk, Sumbudy, Mhsanabary9, RSStockdale, OKBot, Seedbot, Svick, StaticGull, Anchor Link Bot, S2000magician, Randomblue, Driftwood87, Danieltiger45, Dabomb87, Dolphin51, Nergaal, Steve, Athenean, Fscu8014, Clidiere, ClueBot, Trojancowboy, Rumping, Justin W Smith, The Thing That Should Not Be, Plastikspork, Mild Bill Hiccup, Skytopia, Niceguyedc, Jalanpalmer, Luckibrian, Pointillist, DragonBot, LeoFrank, He7d3r, Alejandrocaro35, Sun Creator, Mattspevack, Brews ohare, NuclearWarfare, Jotterbot, Madkaugh, LarryMorseDCOhio, Rylann, ChrisHodgesUK, Thingg, Count Truthstein, Shane-n-phillips, JDPhD, Qwfp, Johnuniq, SoxBot III, RMFan1, Darkicebot, TimothyRias, EdChem, Ultra.Power, Pichpich, Wikiuser100, Chrismacgr, Kal-El-Bot, Weitzhandler, Beach drifter, HOOTmag, Amitahanda, Klundarr, Addbot, Nilesj, Narayansg, Crazysane, Fgnievinski, WardenWalk, Jacraton, Fieldday-sunday, Xforty-enwiki, Bte99, CanadianLinuxUser, Mohamed Magdy, Download, Roux, Favonian, Ozob, Tide rolls, Lightbot, ZedlikBot, Gail, Zorrobot, HerculeBot, Legobot, Luckas-bot, Yobot, OrgasGirl, Timeroot, FUZxxl, PMLawrence, Vini 17bot5, Tempodivalse, AnomieBOT, Rubinbot, Damngoths, Piano non troppo, Kingpin13, Materialscientist, Citation bot, Empro2, Galfam, ArthurBot, LilHelpa, FactSpewer, Xqbot, Mark Bisnaz, Vegpuff, Capricorn42, Flavio Guitian, DSisyphBot, Br77rino, Isheden, Bertvanwijck, GrouchoBot, Omnipaedista, Instantramen92, N419BH, JeanandJane, JonDePlume, Ricklethickets, Semistablesystem, Sesu Prime, FrescoBot, Anna Roy, LucienBOT, Nageh, MathHisSci, Eball, Alxeedo, MathFacts, Haeinous, Musicalvendetta2, Wophopisopenopanoptop, Robo37, Louperibot, Cannolis, Muslim-Researcher, Aturen, Citation bot 1, J. Sketter, Kopiersperre, DrilBot, Pinethicket, Focus, Kiefer.Wolfowitz, Tomcat7, Stpasha, Jschnur, Ezrdr, Beao, December21st2012Freak, Nathan43, Sultan11, Bertina1230, TobeBot, Kingpiko, Math.geek3.1415926, சதீஷ் சிவகுமார், MrX, Suffusion of Yellow, Tbhotch, Dakuton, B3zocdeq8n, RjwilmsiBot, Nathan.C.Heston, TjBot, Dryanm, Jowa fan, Elium2, Whywhenwhohow, EmausBot, Jorgy343, GA bot, Colinbyfleet, 666 Eddie, Arkanoidi, Elementaro, Wham Bam Rock II, Hohho56oy, John Cline, Claudio M Souza, Peteypaws, Fæ, MithrandirAgain, Kiwi128, Dalarocca, Quondum, D.Lazard, Vishnuthelegend, Resprinter123, Aidarzver, Nxtfari, L Kensington, Donner60, Chewings72, Tim Zukas, Tijfo098, ChuispastonBot, DASHBotAV, ResearchRave, Incubheat, ClueBot NG, Jack Greenmaven, Iiii I I I, Matthiaspaul, Hdreuter, O.Koslowski, Iusethis, Artiza11, Gusty55, Helpful Pixie Bot, ZachSchillaci, HMSSolent, Calabe1992, Cenkner, Maonaqua, Walrus068, Vagobot, SRWikis, John Cummings, Haedafae, Samspaki, Mr.miniman67, Thumani Mabwe, Davidiad, Solomon7968, Wiki-art-name, Michael Barera, Souravdas33666, JayEB, Snstrand, Will Gladstone, Brad7777, Remunk, ThatBlokeInTheMask, BattyBot, IkamusumeFan, SupernovaExplosion, Automotiveman, Mharzat123, Jamesbath, JYBot, Dexbot, Boesball, Dushanholt, Thatoneguy617, Frosty, Nicholas Harrison, RandomLittleHelper, N1tkeane, Faizan, Epicgenius, Jose Brox, Mknanda2, Eswasi, Blackbombchu, Ugog Nizdast, Reallilac, W. P. Uzer, Katiewntrs, OccultZone, Stamptrader, Snydergd, Monkbot, Yefeiyao, Qwertyabe12398, Evolutionvisions, Crystallizedcarbon, Jb.haverhill, Loraof, Arvind asia, Ectrum Spanalyser, Robabel41, Some Gadget Geek, Monwolfie, Lúcia D. Coelho, Brockronin, Captaineer, KasparBot, Jeriel123, Bncrump, Acdvorak, SnowFox14 and Anonymous: 841

- **Ceteris paribus** *Source:* https://en.wikipedia.org/wiki/Ceteris_paribus?oldid=681593626 *Contributors:* Matusz, Deb, Michael Hardy, Kku, Kingturtle, Evercat, Radgeek, Topbanana, Banno, Sverdrup, Aetheling, Guy Peters, Dick Bos, Andycjp, Breez, Piotrus, Zfr, Lucidish, DanielCD, Florian Blaschke, El C, Saturnight, Walkiped, Tmh, Grutness, John Quiggin, Arvedui, Saga City, Evil Monkey, Joriki, Velho, OwenX, Woohookitty, LOL, Sburke, Umofomia, Reisio, Mendaliv, Rjwilmsi, Barklund, THE KING, FlaBot, XanaX, FrankTobia, Sceptre, RussBot, Arado, Gaius Cornelius, Bota47, Avraham, Robotico, Sambc, Kungfuadam, SmackBot, Ignotum per Ignotius, Yuriy75, Kintetsubuffalo, SlimJim, Kostmo, Xbxg32000, Scalene, Slife, Cybercobra, Nakon, Richard001, Drphilharmonic, Esrever, Kuru, Gobonobo, Dr.K., MTSbot~enwiki, Dreftymac, Tawkerbot2, OS2Warp, Thomasmeeks, Sdorrance, Gregbard, Tawkerbot4, Thijs!bot, Luna Santin, TheEditrix2, Magioladitis, VoABot II, JNF Tveit, Uncle Dick, Ekkehart, FJPB, DorganBot, Timbabwe, Cpt ricard, Tpb, SieBot, JohnManuel, Quest for Truth, Flyer22, Iloveej, Jaccos, ClueBot, The Thing That Should Not Be, Qwfp, Addbot, Dthompsonza, Numbo3-bot, Luckas-bot, Yobot, Ayrton Prost, Neptune5000, Kingpin13, CasperBraske, Citation bot, GB fan, Butterflybluealias, Ammubhave, Dr Oldekop, Prari, FrescoBot, LucienBOT, Pinethicket, A8UDI, Dinamik-bot, Alph Bot, Osmium192, Palosirkka, Llightex, ClueBot NG, Curb Chain, EkkehartS, Blurrim, Occurring, Skipas750, Loraof, Rohit Gude and Anonymous: 111

- **Scientific modelling** *Source:* https://en.wikipedia.org/wiki/Scientific_modelling?oldid=683116221 *Contributors:* Ronz, Andres, Ike9898, Patrick0Moran, Jon Roland, Mayooranathan, Centrx, Tom harrison, El C, Dalf, Bobo192, Evolauxia, Mdd, Atlant, Velella, Marc A. Dubois, Firsfron, Woohookitty, Linas, Prashanthns, Nobbie, Josh Parris, Sjö, Spencerk, YurikBot, RadioFan, Pseudomonas, Anomalocaris, Grafen, Yahya Abdal-Aziz, ScottyWZ, Erik Sandberg, SmackBot, DCDuring, Cazort, Benjaminevans82, Bluebot, Robth, Mwtoews, FlyHigh, Tktktk, Bjankuloski06en~enwiki, Stwalkerster, Iridescent, Aeternus, CmdrObot, Floridi~enwiki, Pgr94, Gregbard, Cydebot, Al Lemos, Headbomb, West Brom 4ever, Mailseth, Luna Santin, JAnDbot, Stephanhartmannde, Sukratu Barve, MER-C, VoABot II, EagleFan, Squidonius, MartinBot, Anne97432, CommonsDelinker, Erkan Yilmaz, Eliz81, TheSeven, MONODA, Gusfre, Funandtrvl, VolkovBot, AlnoktaBOT, Begewe, Malinaccier, Kilmer-san, Monty845, Pdfpdf, Noiseball, Equilibrioception, Gerakibot, Katonal, Oculi, Sanya3, Sunrise, DancingPhilosopher, Sgagnon, SlackerMom, ClueBot, Bmotoc, Excirial, Brews ohare, Hans Adler, SchreiberBike, Chacen, Dellexxx, Davemody, Gmeltser, SilvonenBot, Addbot, DougsTech, Aktsu, Teles, Yobot, Ptbotgourou, KamikazeBot, DemocraticLuntz, Galoubet, Hommeles, Materialscientist, RibotBOT, MLauba, Spiralforward, Some standardized rigour, FrescoBot, Kdn1982, Wgbh66, JokerXtreme, Emble64, The Stick Man, EmausBot, LÊ TÂN LỘC, Architectchao, Cortes IDS5717C, 超级猪, Wcfios, Rocketrod1960, ClueBot NG, Chilllls, MerllwBot, Helpful Pixie Bot, PhnomPencil, Wingroras, Frosty, Gladtobeherenow, Pamphilia, ModalPeak, BillWhiten, Cmattison387, Aubreybardo, Raad Z Homod, Noyster, WikiTikiTaki 53, SolidPhase, GeneralizationsAreBad, VerdenalH, Canis lupulus, Reshmiiyer and Anonymous: 142

18.9.2 Images

- **File:1_over_x_integral.svg** *Source:* https://upload.wikimedia.org/wikipedia/commons/f/f2/1_over_x_integral.svg *License:* CC BY-SA 3.0 *Contributors:* Own work *Original artist:* Krishnavedala

- **File:3D_Cartesian_Coodinate_Handedness.jpg** *Source:* https://upload.wikimedia.org/wikipedia/commons/b/b2/3D_Cartesian_Coodinate_Handedness.jpg *License:* CC BY-SA 3.0 *Contributors:* Own work *Original artist:* Primalshell

- **File:4octavesAndfrequencies.jpg** *Source:* https://upload.wikimedia.org/wikipedia/commons/e/ee/4octavesAndfrequencies.jpg *License:* CC BY-SA 3.0 *Contributors:* Transferred from en.wikipedia
Original artist: GaulArmstrong. Original uploader was GaulArmstrong at en.wikipedia

- **File:4octavesAndfrequenciesEars.jpg** *Source:* https://upload.wikimedia.org/wikipedia/commons/0/08/4octavesAndfrequenciesEars.jpg *License:* CC BY-SA 3.0 *Contributors:* Transferred from en.wikipedia *Original artist:* GaulArmstrong. Original uploader was GaulArmstrong at en.wikipedia

- **File:Akademia_Ekonomiczna_w_Krakowie_Pawilon_C.JPG** *Source:* https://upload.wikimedia.org/wikipedia/commons/d/d4/Akademia_Ekonomiczna_w_Krakowie_Pawilon_C.JPG *License:* CC BY-SA 3.0 *Contributors:* ? *Original artist:* ?

- **File:Ambox_important.svg** *Source:* https://upload.wikimedia.org/wikipedia/commons/b/b4/Ambox_important.svg *License:* Public domain *Contributors:* Own work, based off of Image:Ambox scales.svg *Original artist:* Dsmurat (talk - contribs)

- **File:Animation_of_exponential_function.gif** *Source:* https://upload.wikimedia.org/wikipedia/commons/0/0b/Animation_of_exponential_function.gif *License:* CC BY-SA 4.0 *Contributors:* Own work *Original artist:* Almacantar

- **File:Anscombe'{}s_quartet_3.svg** *Source:* https://upload.wikimedia.org/wikipedia/commons/e/ec/Anscombe%27s_quartet_3.svg *License:* CC BY-SA 3.0 *Contributors:*

- Anscombe.svg *Original artist:* Anscombe.svg: Schutz

- **File:Aspect-ratio-4x3.svg** *Source:* https://upload.wikimedia.org/wikipedia/commons/d/de/Aspect-ratio-4x3.svg *License:* Public domain *Contributors:* own work, manual SVG coding *Original artist:* Tanya sanderson

- **File:Atmosphere_composition_diagram-en.svg** *Source:* https://upload.wikimedia.org/wikipedia/commons/a/a3/Atmosphere_composition_diagram-en.svg *License:* Public domain *Contributors:* Strategic Plan for the U.S. Climate Change Science Program *Original artist:* Phillpe Rekacewicz

- **File:Benfords_law_illustrated_by_world'{}s_countries_population.png** *Source:* https://upload.wikimedia.org/wikipedia/commons/0/0b/Benfords_law_illustrated_by_world%27s_countries_population.png *License:* CC BY-SA 3.0 *Contributors:* Own work *Original artist:* Jakob.sch

- **File:Logarithm_derivative.svg** *Source:* https://upload.wikimedia.org/wikipedia/commons/5/57/Logarithm_derivative.svg *License:* CC BY-SA 3.0 *Contributors:* Own work *Original artist:* Krishnavedala

- **File:Logarithm_inversefunctiontoexp.svg** *Source:* https://upload.wikimedia.org/wikipedia/commons/4/49/Logarithm_inversefunctiontoexp.svg *License:* CC0 *Contributors:* Own work *Original artist:* Stpasha

- **File:Logarithm_keys.jpg** *Source:* https://upload.wikimedia.org/wikipedia/commons/8/88/Logarithm_keys.jpg *License:* CC BY 2.0 *Contributors:* 100502-1150494 *Original artist:* Waifer X

- **File:Logarithm_visualization_tree.svg** *Source:* https://upload.wikimedia.org/wikipedia/commons/6/61/Logarithm_visualization_tree.svg *License:* CC BY 3.0 *Contributors:* Own work *Original artist:* George Snyder

- **File:Logarithms_Britannica_1797.png** *Source:* https://upload.wikimedia.org/wikipedia/commons/8/82/Logarithms_Britannica_1797.png *License:* Public domain *Contributors:* Britannica 1797 vol. 10 p.119 via google book search *Original artist:* Colin Macfarquhar (editor)

- **File:Loudspeaker.svg** *Source:* https://upload.wikimedia.org/wikipedia/commons/8/8a/Loudspeaker.svg *License:* Public domain *Contributors:* New version of Image:Loudspeaker.png, by AzaToth and compressed by Hautala *Original artist:* Nethac DIU, waves corrected by Zoid

- **File:MathModel.svg** *Source:* https://upload.wikimedia.org/wikipedia/commons/f/f2/MathModel.svg *License:* Public domain *Contributors:* Own work, inspired by Bertuglia & Vaio 2005 *Original artist:* Tomaschwutz

- **File:Merge-arrow.svg** *Source:* https://upload.wikimedia.org/wikipedia/commons/a/aa/Merge-arrow.svg *License:* Public domain *Contributors:* ? *Original artist:* ?

- **File:Mergefrom.svg** *Source:* https://upload.wikimedia.org/wikipedia/commons/0/0f/Mergefrom.svg *License:* Public domain *Contributors:* ? *Original artist:* ?

- **File:Modeling_and_Simulation_Integrated_Use.jpg** *Source:* https://upload.wikimedia.org/wikipedia/commons/c/c0/Modeling_and_Simul Integrated_Use.jpg *License:* Public domain *Contributors:* ? *Original artist:* ?

- **File:Natural_logarithm_integral.svg** *Source:* https://upload.wikimedia.org/wikipedia/commons/d/df/Natural_logarithm_integral.svg *License:* Public domain *Contributors:*

- Log-pole-x.svg *Original artist:* Log-pole-x.svg: Wojciech Muła

- **File:Natural_logarithm_product_formula_proven_geometrically.svg** *Source:* https://upload.wikimedia.org/wikipedia/commons/9/9b/Na logarithm_product_formula_proven_geometrically.svg *License:* Public domain *Contributors:*

- Natural_logarithm_integral.svg *Original artist:* Natural_logarithm_integral.svg: *Log-pole-x.svg: Wojciech Muła

- **File:NautilusCutawayLogarithmicSpiral.jpg** *Source:* https://upload.wikimedia.org/wikipedia/commons/0/08/NautilusCutawayLogarithm jpg *License:* CC BY-SA 3.0 *Contributors:* ? *Original artist:* ?

- **File:Nuvola_apps_edu_mathematics_blue-p.svg** *Source:* https://upload.wikimedia.org/wikipedia/commons/3/3e/Nuvola_apps_edu_math blue-p.svg *License:* GPL *Contributors:* Derivative work from Image:Nuvola apps edu mathematics.png and Image:Nuvola apps edu mathematics-p.svg *Original artist:* David Vignoni (original icon); Flamurai (SVG convertion); bayo (color)

- **File:PDF-log_normal_distributions.svg** *Source:* https://upload.wikimedia.org/wikipedia/commons/a/ae/PDF-log_normal_distributions.svg *License:* CC0 *Contributors:* Own work *Original artist:* Krishnavedala

- **File:PSM_V36_D057_Hyperbolas_produced_by_interference_of_waves.jpg** *Source:* https://upload.wikimedia.org/wikipedia/commons/c/c0/PSM_V36_D057_Hyperbolas_produced_by_interference_of_waves.jpg *License:* Public domain *Contributors:* Popular Science Monthly Volume 36 *Original artist:* Unknown

- **File:People_icon.svg** *Source:* https://upload.wikimedia.org/wikipedia/commons/3/37/People_icon.svg *License:* CC0 *Contributors:* OpenClipart *Original artist:* OpenClipart

- **File:Percent_18e.svg** *Source:* https://upload.wikimedia.org/wikipedia/commons/f/f9/Percent_18e.svg *License:* Public domain *Contributors:* Created with the DejaVu Sans font and Inkscape. *Original artist:* Farmer Jan and bdesham

- **File:Polynomialdeg2.svg** *Source:* https://upload.wikimedia.org/wikipedia/commons/f/f8/Polynomialdeg2.svg *License:* Public domain *Contributors:* <a href='//commons.wikimedia.org/wiki/File:Polynomialdeg2.png' class='image'><img alt='Polynomialdeg2.png' src='https://upload.wikimedia.org/wikipedia/commons/1/14/Polynomialdeg2.png' width='233' height='179' data-file-width='233' data-file-height='179' /></a> *Original artist:* Original hand-drawn version: N.Mori

- **File:Portal-puzzle.svg** *Source:* https://upload.wikimedia.org/wikipedia/en/f/fd/Portal-puzzle.svg *License:* Public domain *Contributors:* ? *Original artist:* ?

- **File:Proportional_variables.svg** *Source:* https://upload.wikimedia.org/wikipedia/commons/0/03/Proportional_variables.svg *License:* CC0 *Contributors:* Own work *Original artist:* Krishnavedala

- **File:Question_book-new.svg** *Source:* https://upload.wikimedia.org/wikipedia/en/9/99/Question_book-new.svg *License:* Cc-by-sa-3.0 *Contributors:* Created from scratch in Adobe Illustrator. Based on Image:Question book.png created by User:Equazcion *Original artist:* Tkgd2007

- **File:Rechte-hand-regel.jpg** *Source:* https://upload.wikimedia.org/wikipedia/commons/7/79/Rechte-hand-regel.jpg *License:* CC-BY-SA-3.0 *Contributors:* Own work *Original artist:* Abdull

- **File:Reciprocal_integral.svg** *Source:* https://upload.wikimedia.org/wikipedia/commons/c/c8/Reciprocal_integral.svg *License:* CC0 *Contributors:* Own work *Original artist:* User:Dcoetzee

18.9.3 Content license

www.ingramcontent.com/pod-product-compliance
Lightning Source LLC
Chambersburg PA
CBHW081442170526
45166CB00008B/2289